Modern China: A Very Short Introduction

VERY SHORT INTRODUCTIONS are for anyone wanting a stimulating
and accessible way into a new subject. They are written by experts, and
have been translated into more than 40 different languages.

The series began in 1995, and now covers a wide variety of topics in every
discipline. The VSI library now contains over 450 volumes—a Very Short
Introduction to everything from Psychology and Philosophy of Science to
American History and Relativity—and continues to grow in every subject
area.

Very Short Introductions available now:

Available soon:

SLANG Jonathon Green

EARTH SYSTEM SCIENCE Tim Lenton

THE WELFARE STATE David Garland

CRYSTALLOGRAPHY A. M. Glazer

SHAKESPEARE'S COMEDIES
Bart van Es

For more information visit our website

www.oup.com/vsi/

Rana Mitter

MODERN CHINA

A Very Short Introduction
SECOND EDITION

OXFORD
UNIVERSITY PRESS

Great Clarendon Street, Oxford, OX2 6DP,
United Kingdom

Oxford University Press is a department of the University of Oxford.
It furthers the University's objective of excellence in research, scholarship,
and education by publishing worldwide. Oxford is a registered trade mark of
Oxford University Press in the UK and in certain other countries

First edition published 2008
Second edition published 2016

Impression: 12

Published in the United States of America by Oxford University Press
198 Madison Avenue, New York, NY 10016, United States of America

British Library Cataloguing in Publication Data

Data available

Library of Congress Control Number: 2015949960

ISBN 978-0-19-875370-4

Printed in Great Britain by
Ashford Colour Press Ltd., Gosport, Hampshire.

Contents

Acknowledgements

I am very grateful to all those at Oxford University Press who commissioned this *Very Short Introduction* and saw it through its various stages of life: Marsha Filion, Luciana O'Flaherty, Deborah Protheroe, and James Thompson. Writing this book made copious reading necessary. I could not have easily made the time to do that reading and reflection without teaching relief funded by the generous grant of a Philip Leverhulme Prize by the Leverhulme Trust, for which I am immensely grateful. On-the-ground observations were made possible by the kind award of a place on the British Academy–Chinese Academy of Social Sciences exchange scheme. I also owe thanks to the anonymous reviewers who gave valuable comments both at proposal and manuscript stage. Colleagues and friends contributed in many ways to the book, but I must single out Graham Hutchings and Neil Pyper, who patiently read and commented on the whole of a very rough draft with wit and copious good sense. For the second edition, I have benefited from generous and detailed anonymous readers who have saved me from many errors, as well as excellent editing from Emma Ma and Jenny Nugee. I have also had the constant support of my parents and Pamina, and I have also had the joy of Malavika's arrival between the first and second editions. Katharine read and commented on the entire manuscript, and offered support in countless other ways. This book is dedicated to her.

Pronunciation

This book uses the *pinyin* system of romanization of Chinese. Very approximately indeed, the transliterations that cause most problems for English-speakers are the following sounds:

    c—pronounced 'ts'
    x—pronounced 'sh'
    q—pronounced 'ch'

And the sounds *chi, zhi, ri, si, shi,* and *zi* are pronounced as if the 'i' sound is an 'rr'—so 'chr', 'zhr', and so forth.

List of illustrations

Chapter 1
What is modern China?

> It is impossible to do other than assent to the unanimous verdict
> that China has at length come to the hour of her destiny...The
> contempt for foreigners is a thing of the past...Even in remote
> places we have found the new spirit—its evidence, strangely
> enough, the almost universal desire to learn English...as
> knowledge of English is held to be the way to advancement, the key
> to a knowledge of the science and art, the philosophy and policy,
> of the West.

This assessment comes from the book *New China* by W. Y. Fullerton
and C. E. Wilson. In the fourth decade of China's era of
'reform and opening-up' (*gaige kaifang*), at last the clichés of
the old Maoist era—the Chinese as worker ants mouthing
xenophobic anti-imperialist slogans while all dressed in blue serge
boiler-suits—have given way to impressions of a country whose
cities are full of skyscrapers, whose rural areas are being transformed
by new forms of land ownership and a massive rise in migrant
labour, and whose population is keen to engage with the outside
world after years of isolation. Fullerton and Wilson's observation
that China is reaching the 'hour of her destiny', and that a
significant part of the population are learning English as one way
to fulfil that destiny, seems a reasonable comment on a China that
is clearly very different from the one ruled a generation ago by
Chairman Mao.

However, Fullerton and Wilson did not pen their observations having landed back at Kennedy or Heathrow airports on one of the many Air China 777s that ferry thousands of travellers daily between China and the West. They wrote their book a full century ago, and their reflections on what they subtitle 'a story of modern travel' came at what, in retrospect, is a particularly poignant moment in China's history: the year 1910. The China they portray was lively, even optimistic, and very much engaged with the outside world. Yet within a year, the Qing dynasty, the last Chinese imperial house which still ruled the country that Fullerton and Wilson saw, had fallen. The revolution of October 1911 finally brought the two-thousand-year tradition of imperial rule in China to an end, making way for a Republic. That Republic, Asia's first, would collapse less than 40 years later, and would be succeeded in turn by a People's Republic whose own form would change over decades as it struggled to define what 'modern' China was. It is a sign of how long it has taken for China to define its own vision of modernization that a travel description from the early 20th century can still have resonance in the early 21st.

China is the world's most populous country, with over 1.3 billion inhabitants as of 2013. Its economy grew in the first decade of this century by an average of around 10 per cent a year. It is seeking a regional and global role, with a new political and economic presence in Africa, Latin America, and the Middle East, and has taken frequent steps to portray itself as a responsible member of the world community, playing a role in troublesome areas such as North Korea where the West has little sway. The 2008 Beijing Olympics marked the 'coming-out' of China as an integrated member of the world community of nations, the acme of the 'peaceful rise' which it has been engineering since the mid 1990s. The term 'peaceful rise' (*heping jueqi*) itself, associated with the political thinker Zheng Bijian, was thought by Chinese ideologists to be too confrontational, and has been replaced with the term 'peaceful development'. The idea remains the same, however: that

China is finally gaining the role as a regional and global power that it lost in the mid-19th century.

Everywhere one goes in China, there are signs of change. Significant areas of western China have been flooded to make way for the massive Three Gorges Dam on the Yangtze River. Their former inhabitants are being relocated and urbanized as China moves away from its traditional agricultural past. In the cities, Baidu, a home-grown Chinese internet search engine, dominates the market which is held by the worldwide brand leader, Google, in most other countries. Beneath China's strict censorship laws lies a 'grey zone' of cultural production: from underground movies criticizing the Cultural Revolution to pornography, cultural rebels find ways to make their views known.

China is now a major actor in world markets. Through much of the early 2000s, China's burgeoning exports led to concerns in the USA and EU about the Chinese trade surplus. The West was also concerned about the strength of the Chinese currency (the yuan or renminbi) against the dollar, with the Americans and French frequently lobbying the People's Bank of China to revalue it upwards. The Chinese current account surplus also means that it has had cash to spend on investments around the world, from the US, to Africa, to Russia.

Yet China is also undertaking one of the most precarious balancing acts in world history. While the country has the fastest-growing major economy in the world, it is also becoming one of the globe's most unequal societies, even while its policies lift millions out of poverty (See Figure 1). For the rural and urban poor, higher-level health care and education are often available only to those who can pay for them. China is also in the grip of a resource and environmental crisis. All across China, power blackouts regularly interrupt industrial production. Globally, the country must scramble for energy and mineral resources. Environmental degradation forces bicyclists to wear smog masks and has rendered the Yangtze

1. Migrant labourers without local residence permits are a common sight on Chinese building sites. Their work underpins the futuristic skylines in cities such as Shanghai and Beijing.

dolphin extinct. As climate change accelerates, China has become the world's largest emitter of carbon dioxide into the atmosphere. China continues to maintain a one-party authoritarian state and heavily constrains political dissent; yet every year, there are thousands of demonstrations against official policies and practice, some of them violent. Corruption also runs rife.

There are significant differences between China at the start of the 20th century, and the early 21st. The China of a century ago was the victim of Western and Japanese imperialism, in danger, in the phrase of the time, of 'being carved up like a melon' by the foreign powers. It was a weak and vulnerable state. Today's China, while it has deep frictions and fault lines, is a much stronger entity. Yet the similarities between China now and China a hundred years ago are startling also: political instability, economic and social crisis, and the need for China to find a role in a world dominated, even if less so than in the post-Cold War moment, by the West.

Chinese leaders, who are acutely conscious of history in a way that has been less true of the American and British governing classes in recent years, would also note that the seemingly moribund Qing dynasty had begun to modernize impressively fast in the early years of the 20th century. Nonetheless, it collapsed, as did most of its successor regimes in the following four decades. It is the earnest intention of the rulers of the People's Republic of China that this fate should never happen to them. To understand their fears and concerns, and to understand China in its own terms, the China of today can only be understood in its historical and global context. That is what this book tries to do, explaining the reasons that modern China looks the way that it does.

Overall, the book hopes to give a picture of China that reflects three main viewpoints. First, rather than being a closed society, China has almost always been a society open to outside influence, and 'Chinese' culture and society cannot be understood in isolation from the outside world. In other words, China cannot be treated as a special case of an isolated society, but rather as part of a changing regional and global culture. Second, it is too simple to say that China has moved from a 'traditional' past to a 'modern' present. Rather, the modern China we see today is a complex mixture of indigenous social influences and customs and external influences, often, but not always, from the West. Society did not change overnight in 1912 with the abdication of the last emperor, or in 1949 with the Communist revolution, but neither is the modern China of today essentially the same as when the emperors were on the throne a hundred or two hundred years ago. Third, our understanding of how modern China has developed should not come only through following elite politics and leaders and their conflicts. Instead, we should look at continuities as well as changes in how the Chinese have come to modernity, and the impact of change on society and culture as a whole.

What does it mean to be Chinese?

A hundred years ago and today, an important question remains: What *is* modern China? To come to an answer, we need to spend a little time investigating both terms—*China* and *modern*.

'China' today generally refers to the People's Republic of China, the state that was established in 1949 after the victory of the Chinese Communist Party (CCP) under the leadership of Chairman Mao Zedong. That state essentially covers the same territory as the Chinese empire under the last imperial dynasty, the Qing (1644–1911), which extended its reach to the west and north of the lands that earlier dynasties had controlled. (The modern state, however, has a firm grasp on Tibet, does not lay claim to Outer Mongolia, which separated from China in 1911, and in practice does not control Taiwan.) However, this continuity of geography conceals the reality that China has changed shape over the centuries, and continues to do so even now. About 2,500 years ago, a group of independent states that were in conflict with one another existed in the heartland of what we now call 'China'; literature and history from this period is recognizably Chinese, readable by those today who take the trouble to learn the classical form of the language. From 221 BCE, successive emperors and dynasties united these states, leading to a succession of dynasties that created China's classical civilization: the Han, the Tang, the Song, the Yuan, the Ming, and the Qing among them. They created a civilization in which art, literature, statecraft, medicine, and technology all thrived.

However, the term 'China', or the term *Zhongguo* ('middle kingdom', the current Chinese word for 'China'), was not how the people of those eras would have thought of themselves. The idea of being 'Chinese' in the sense that we understand it, as either national or ethnic identity, is a product of the 19th century. Yet there clearly was a shared sense of what we might call 'Chineseness' between these people, which outlasted the rise and fall of dynasties. What made up that identity? Most people identified themselves with the ruling dynasty itself, as 'people of the Ming' or 'people of the Qing'. But what lay behind this naming? How did one qualify as a 'person of the Ming'?

Over the centuries, there has been a variety of shared attributes that have brought together the communities we know as 'the Chinese'. From early on, Chinese society was settled and agricultural, in contrast with the nomad societies such as the Manchus, Mongols, and Jurchen with which it periodically came into contact. Features of that society, such as irrigation, have also been prominent throughout Chinese history. The size of the Chinese population has always dwarfed its neighbours, and that population has increased with territorial growth over the centuries as well. In very early China, the landmass was occupied by a variety of peoples, but from 221 BCE, after the unification by the Qin dynasty, dominance remained with a people whom we recognize as Chinese (often called 'Han' Chinese after the next dynasty).

But why did the *Chinese* think of themselves as Chinese? Broadly, shared identity came from shared rituals. For more than 2,000 years, a set of social and political assumptions, which found their origins in the ideas of Confucius, a thinker of the 6th century BCE, shaped Chinese statecraft and everyday behaviour. By adopting these norms, people of any grouping could become 'people of the dynasty'—that is, Chinese.

Confucianism is sometimes termed a religion, but it is really more of an ethical system, or system of norms. In its all-pervasiveness *and* its flexibility and adaptability to circumstances, it is somewhat analogous to the role of Judaeo-Christian norms in Western societies, where even those who dispute or reject those norms still find themselves shaped by them, consciously or not. Confucianism is based on ideas of mutual obligation, maintenance of hierarchies, a belief in self-development, education, and improvement, and above all, an ordered society. It abhors violence and tends to look down on profit-making, though it is not wholly opposed to it. The ultimate ideal was to become sufficiently wise to attain the status of 'sage' (*sheng*), but one should at least strive to become a *junzi*, often translated as 'gentleman', but perhaps best thought of as meaning 'a person of integrity'. Confucius looked back to the Zhou dynasty, a supposed 'golden age' which was long-past even during his lifetime, and which set a desirable (but perhaps unattainable) standard for the present day.

Confucius's opinions did not emerge from thin air: he lived during the period of the Warring States, a violent era whose values appalled him, and which fuelled his concern with order and stability. Nor was he the only thinker to shape early China: unlike Confucius and Mencius, who believed in the essential good nature of human beings, Xunzi believed that humans were essentially evil; and Han Feizi went further to argue that only a system of strict laws and harsh punishments, not ethical codes, could restrain people from doing wrong. This period, the 5th century BCE, was one of profound crisis in the territory we know now as China, but ironically, it led to an unmatched excellence in the cultural and intellectual atmosphere of the time, just as the crisis of 5th-century BCE Greece led to an extraordinary outpouring of drama and philosophy. Despite the intellectual ferment of the time, it was Confucius's thought that became most acceptable in Chinese statecraft, although his ideas were adapted, often beyond recognition, by the statesmen and thinkers who drew on his writings over the centuries. Nonetheless, throughout that period, assumptions from Confucianism persisted.

The premodern Chinese had a clear idea of a difference between themselves and other groupings, not least because there were frequent attacks by and on the neighbours. During two of China's greatest dynasties, the Yuan and the Qing, the country was ruled by ethnic non-Chinese (Mongols and Manchus respectively). However, the remarkable resilience of the Chinese system of statecraft meant that these occupiers soon adapted themselves to Chinese norms of governance, something that marked these invaders out from the Western imperialists, who did nothing of the sort. The assimilation was not total. The Qing aristocracy maintained a complex system of Manchu elite identity during their centuries of power: Manchus were organized in 'banners' (groupings based on their military nomadic past), and Manchu women did not bind their feet. But overall, the rituals and assumptions of Confucian ethics and norms still pervaded society: Qing China was at core a Chinese, not a Manchu, society (See Figure 2).

The 19th century saw a profound change in Chinese self-perception. For centuries, the empire had been termed *tianxia*, literally and poetically rendered as 'all under heaven'. This did not mean that premodern Chinese did not recognize that there were lands or peoples that were not their own—they certainly did—but that the empire contained all those who mattered, and its border was flexible, although not infinitely elastic. (The Treaty of Nerchinsk, signed in 1689, drew up the border which still exists today between China and Russia; clearly Qing China did not lack a sense of territoriality.)

But the arrival of Western imperialism forced China, for the first time, to think of itself as part of an international *system*. The arrival of European political thought brought to China the idea of the nation-state, and many Chinese came to terms with the fact that the old China was gone, and that the new one would need to assert its place in the hierarchy of nations. That struggle is still with us today.

Yet the modern People's Republic does not contain the whole of China, or China's worlds, within it. Taiwan provides an alternative, lively, and democratic vision of what 'Chinese culture' is; so does

2. A wealthy Chinese woman in the early 20th century, with expensive clothes and bound feet.

Hong Kong. Then there are diaspora Chinese: the 'overseas Chinese' who shape societies such as Singapore and whose communities are found on all inhabited continents.

China is a continent, not just a country. It is a series of identities, some shared, some differentiated, and some contradictory: modern, Confucian, authoritarian, democratic, free, and restrained. Above all, China is a plural noun.

What is modern?

Frequently, 'modern' is used as shorthand for 'recent'—so a study of 'modern' China would refer to its history over the last century or so. This book, however, will use a more specific definition of 'modern', because by doing so, it can get to the heart of some of the biggest questions that continue to face China today—the questions of what sort of society and culture it is, and wishes to become.

First, though, there are certain ways *not* to think about 'modern' China. When trying to define the way in which China has changed since the 19th century, it is possible to fall into one of two overly broad explanations.

The first explanation was more common a generation ago, when Mao was in power and China seemed utterly to have changed its political and social system. This argument followed the CCP's rhetoric of a 'new China' (although, as the quotation at the start of this chapter shows, this was not the first nor last usage of the term 'new China'): that the old, 'feudal', 'traditional', and 'semicolonial' China, a world of cruel social hierarchies, foot-binding, torture, and poverty, had been finally brushed aside for a more egalitarian, industrial, and just China.

The second explanation, common in the early 20th century, but banished for a while after 1949, has become commonplace again today. This argument is that China has not essentially changed.

Even figures such as Mao and Deng Xiaoping (the reformist leader of the 1980s), despite their coating of communist ideology and mass mobilization politics, were essentially 'emperors' reverting to type. In the countryside today, traditional superstitions, religions (such as the Falun Gong cult banned by the party), and hierarchies reign supreme, just as they have done for hundreds of years. Overall, China remains a Confucian, hierarchical society with an ostensibly communist brand name on top.

These views are wrong. China is a profoundly modern society; but the way in which its modernity has been manifested is indelibly shaped by the legacy of its premodern (a term preferable to 'traditional') past. Not that the premodern past was ever monolithic or static: China changed immeasurably over hundreds of years, developing a bureaucracy, science, and technology (the invention of gunpowder, clocks, and the compass), a highly commercialized economy (from around 1000 onwards), and a diverse syncretic religious culture.

The similarity in many developments in Europe and China in the period 1000 to around 1800 should not, however, conceal the fact that imperial China and early modern Europe also *differed* widely in their assumptions and mindsets. The development of modernity in the Western world was underpinned by a set of assertions, many of which are still powerful today, about the organization of society. Most central was the idea of 'progress' as the driving force in human affairs. Philosophers such as Descartes and Hegel ascribed to modernity a rationality and teleology, an overarching narrative, that suggested that the world was moving in a particular direction—and that that direction, overall, was a positive one. There were several drivers of progress. One was the idea that dynamic change was a good thing in its own right: in premodern societies, the force of change was often feared as destructive, but the modern mindset welcomed it. In particular, an acceptance and enthusiasm for progress through economic growth, and later, industrial growth, became central to the development of a modern society. Particularly in the formulation of the Enlightenment of the 18th century, the idea of rationality, the ability to make choices and decisions in a predictable, scientific way, also became crucial to the ordering of a modern society.

Modernity also altered the way in which members of society thought of themselves. Society was secularized: modernity was not necessarily hostile to religion, but religion was confined to a defined space within society, rather than penetrating through it. The individual self, able to reason, was now at the centre of the modern world. At the same time, the traditional bonds that the self had to the wider community were broken down; modern societies did not support the old feudal hierarchies of status and bondage, but rather, broke them down in favour of equality, or at any rate, a non-hierarchical model of society.

Above all, societies are modern in large part because they perceive themselves as being so: self-awareness ('enlightenment') is central to modernity and the identities that emerge from it, such as nationhood. This has led the West, in particular, to draw far too strong a distinction between its own 'modern' values and those elsewhere in the world. China, for instance, showed many features over thousands of years that shared assumptions of modernity long before the West had a significant impact there. China used a system of examinations for entry to the bureaucracy from the 10th century CE, a clearly rational and ordered way of trying to choose a power elite, at a time when religious decrees and brute force were doing the same job in much of Europe. At the same time, China started to develop an integrated and powerful commercial economy, with cash crops taking the place of subsistence farming. It is clear that many aspects of 'modernity' were visible earlier and more clearly in China than in Europe.

Among the most powerful elements of modern thought in Europe was its ability to maintain the idea that its own genesis and construction were profoundly different from those of other societies. In part, this was because of a desire to create a clear distinction between Western European politics and that of other societies, particularly in the 19th century, when imperialist ideology became important. Yet in many ways, the attributes of modernity—particularly self-awareness and its associated sense of anti-hierarchy—were drawn from a pre-existing religious tradition, in which birth and rebirth were crucial. While Christianity was clearly one source of this concept (having also provided the cultural grounding for the teleology of progress that underlies classic modernity), the ideas of enlightenment (not, however, expressed as rebirth) and self-awareness emerged much earlier as part of Buddhist thought, and in later centuries were developed within another path defined by Islam. The most strongly Eurocentric understandings of modernity have found it hard to acknowledge its cross-cultural roots; yet they are there.

But all the same, China before the mid-19th century did not share certain key assumptions of the emerging elites of Europe in the 16th to 19th centuries. China did not, during that time, develop powerful political movements that believed in flattening hierarchies: in the Confucian world, 'all men within the four seas' might be 'brothers', but 'all men' were not equal. Chinese thinkers did not stress the individuated self as a positive good in contrast to the collective, although there was a clear idea of personal development to become a 'gentleman' or 'sage'. Nor, overall, did it make the idea of a teleology of forward progress central to the way it viewed the world: rather, history was an attempt to recapture the lost golden age of the Zhou and ways of the ancients, and rather than praising innovation and dynamic change in its own right, premodern China developed highly sophisticated technology and statecraft while stressing the importance of past precedent, and of order. As for economic growth, while it would be too strong to say that Confucian thought wholly disapproved of trade (the Ming and Qing certainly saw a comfortable accommodation by the state with the idea of commerce), the concept of economic progress as a good in its own right was not as central to the premodern Chinese mindset as it was to the type of modernity that emerged in Europe.

These assumptions mark a profound difference from China's experience in the contemporary era. Since the early 20th century at least, China's governments and elite thinkers have accepted most of the tenets of modernity, even while vehemently opposing the Western and Japanese imperialism which forced those ideas into China. As we will see (Chapters 2 and 3), the Communist and Nationalist governments that dominated China in the 20th century both declared that China was progressing towards the future; that a new, dynamic culture was needed to take it there; that hierarchies needed to be broken down, not preserved; and that while order was important, economic growth was the only way to make China rich and strong. Most notably, China's leaders were much more fiercely and uncompromisingly modern in their assumptions than many of their contemporaries in India or Japan in the early 20th century: as Chapter 3 suggests, the 'May Fourth Movement' of the 1910s was far more eager to reject China's Confucian past completely than figures in India, such as Gandhi, were to reject that society's past.

But at the same time, there is a chimerical element to the quest for modernity. Modernity keeps changing, and Chinese conceptions of it change as well: the modernity of the 'self-strengtheners' who sought to adapt Western technology in the late Qing is not the same as that of the radicals who declared a 'new culture' in the 1910s, nor of the Nationalists and Communists whose primary goal was to find a stable, modern identity for the Chinese state and people. Even today, the question of what a modern China looks like is in flux. At the same time, China's new-found strength means that it is in a much better position than ever before to project aspects of its own model of modernity back into the wider global definition of the term.

With very few exceptions, all of the warring factions that vied over China's future in the 20th century were 'modern', not just in the sense of being 'recent', but in their rejection or adaptation of the Confucian norms of the past, and their embrace of a new set of norms that were derived from outside, but which were adapted to make 'Chinese' and 'modern' compatible, rather than terms which seemed to be in opposition to one another. Although they violated their own rhetoric on countless occasions, China's rulers in the 20th century—and the 21st—have sought to create a nation-state with an equal, self-aware citizen body. This is a distinctly modern goal. The rest of the book will seek to assess how successful they have been in achieving it.

Chapter 2
The old order and the new

A typical characterization of China's past, often put forward by Chinese modernizers in the 20th century, is that late imperial China was a corrupt, 'feudal' mess that was held back by unchanging, conservative, Confucian thought. Yet the imperial Chinese state, while underpinned by ideas of order and hierarchy, was also driven by a sense of mutual obligation between different groups in society and gave rise to an ever-changing and highly dynamic political and social culture, although this had collapsed in many important respects by the early 20th century.

However, Western political influence did change China profoundly in the late 19th century in the wake of the Opium Wars, when concepts such as nationalism and Social Darwinism became hugely influential on a generation of Chinese who felt that their country was now vulnerable to the outside world. Japan also became a conduit for importing the new modern modes of thought. In 1868, the revolution known as the Meiji Restoration began, turning Japan in just a couple of decades from a feudal state run by a warrior aristocracy to a modernizing, industrial empire. Among these modern political concepts that energized debate in the very last days of the Qing dynasty were the ideas of a constitution, parliamentary government, citizenship rather than subjecthood, and reorientation of China into an international system of nation-states. Even though the dynasty itself fell, these concepts would shape all political discussion in the 20th century, and are indeed profoundly important today.

The early 20th century was a time of great political distress in China, but also opened up unimagined new vistas for generations of Chinese at all levels of society: rural girls who became factory workers, farmers who became Communist activists, and middle-class students who learned about Japan, Europe, and America at first hand. The imperial system collapsed in the revolution of 1911, and in its place a republic was established. The political atmosphere of the time allowed fierce debates on nationalism, socialism, and feminism, among other ideas, although the Confucian influences and preconceptions still continued to shape everyday life. Civil war and the dominance of imperialist powers in China, however, prevented participatory politics from taking hold, even after the Northern Expedition of 1926–8, which brought the Nationalists to power under Chiang Kaishek, the successor to the veteran Nationalist leader Sun Yatsen. This chapter charts the first stage of that long journey towards a modern mass politics.

The age of gold

The English civil war and the American and French revolutions had a profound influence on the relationship between Western governments and their people. The modernity of those systems lies in certain assumptions: that government should be representative of the people, and that the people have inherent rights to the choice and policies of their government; that government should act rationally and to the greatest benefit of all; and that citizenship, membership in a national body, should be granted on the basis of equality and not assigned by a hierarchy derived from any irrational, arbitrary, or other-worldly source.

It is often assumed that non-European societies such as China shared few, if any, of these assumptions. Certainly some of them ran counter to Confucian norms, yet as we shall see, the question of how the people might be governed justly was critical to many Chinese political discussions. Central to the assumptions of emperors and their officials were Confucian ideas about the makeup of the state. In these ideas, hierarchy was not only present but essential: the body politic was held to be a metaphorical extension of the family; just as sons should obey fathers and wives should obey husbands, so subjects should obey their rulers. The people did not have inherent rights as individuals or even as a collective body.

However, it would be wrong to think that this made Chinese governance arbitrary, irrational, or despotic. A good ruler in the Confucian world was not at liberty simply to do as he pleased. The people were in his charge, and cruel or unfair behaviour towards them would result in his losing the 'mandate of heaven'. Confucius and philosophers who had followed his tradition, such as Mencius, stressed that attendance to the welfare of the people was a primary task for any ruler. The great dynasties of imperial China certainly paid attention to questions of welfare; for example, removing tax burdens on areas where flooding had destroyed crops, and on occasion trying (not very successfully) to maintain 'ever-full granaries' that would hold food reserves for distribution in times of hardship. Government was primarily concerned with the maintenance of order, but to do so, it was clear that the people had to feel that laws were applied fairly and equitably. At times of turmoil, the system became corrupt and dysfunctional, but during its periods of confidence and prosperity, such as the 18th century, the system was one of the world's most successful empires.

Chen Hongmou (1696–1771) was one of the most prominent administrators and thinkers on statecraft of the High Qing, the period during the 18th century when China seemed peaceful and prosperous, a prime example of a well-run empire that looked to the good of its people. The assumptions and contradictions in Chen's writings show clearly the way in which Confucian principle and the realities of governance created an effective, though not always consistent, style of government (and in those contradictions, no great difference from the compromises of Western governments). Simultaneously, Chen and his contemporaries in the Qing bureaucracy showed a commitment to the traditional Chinese patriarchy, yet Chen also declared that 'heavenly goodness' (*tianliang*) lay in all people, even 'petty commoners', lowly '*yamen* (local magistrate's office) clerks', and even those who were not ethnically Han Chinese. Chen and his contemporaries also put into place policies that encouraged social mobility and popular education (including literacy for women), as well as merchant enterprises: none of the latter are popularly associated with 'Confucian' thought, yet Chen advocated them with no sense of violation of the norms for a decent and well-ordered society.

Wei Yuan (1794–1856), one of the most well-known thinkers of the late Qing era, thought broadly about the nature of political participation. He was a Qing loyalist, but strongly argued that the dynasty needed to reform its administration if it were to cope with the threat from overseas. While he never came anywhere close to advocating that the ordinary population of China should take part in their own governance, he wrote extensively about the danger of political sterility that could come from restricting both the numbers and scope for argument of those who could critique power-holders: 'There is no single doctrine which is absolutely correct, and no single person who is absolutely good.' Wei argued for a competition in ideas, that would enable the ruler to choose between rival proposals, and in 1826 made his own contribution to that discussion by publishing the 'Collected Essays on Statecraft'. Wei Yuan, however, did not want to widen political participation so as to reduce the role of the state, but rather to strengthen it.

The new world

If the 18th century was a broadly successful one for China, the 19th saw the Qing dynasty disintegrate under a series of crises both internal and external. The most obvious trigger for collapse was the arrival of the Western imperial powers demanding that China open itself up to their trading demands. But the Western impact alone was not enough to bring down the Qing. The grave internal stresses and strains manifested themselves in the seemingly separate effects of Western imperialism, with both sets of crises feeding off one another.

Internally, the rapid expansion of the size of the Qing empire had led to problems, as the bureaucracy did not increase to match its new responsibilities. Tax collection became more difficult, and increasingly corrupt. Between 1600 and 1800, the size of the population doubled to some 350 million; the number of people who were poor and dissatisfied increased also. Regardless of the Western intrusion into China, one can see in the late Qing the signs of imperial overstretch that had also eventually doomed China's previous dynasties.

Nonetheless, the arrival of European imperialism had effects that had simply not been relevant for the earlier dynasties. The development of the East India Company by the British meant that large quantities of opium being produced in Bengal now needed a market. The Chinese government, after some debate, banned the sale of opium within China, alarmed at its popularity and addictiveness. The British government, newly concerned with empire in Asia, took the ban (and the destruction of British-owned opium in Canton harbour) as a provocation. The first Opium War of 1839–42 saw the Qing government defeated, and forced to concede what would be one of a long list of treaties with foreigners made under duress, and remembered by generations of Chinese as 'unequal'. Between the mid-19th and the mid-20th centuries, Chinese governments were never wholly in control of their own territory. Foreigners under treaty rights had 'extraterritoriality' (that is, they were not subject to Chinese law); a whole series of 'treaty ports' were established in which foreigners had new trading rights (and some places, such as Hong Kong, were fully colonized);

and new and disruptive influences, notably Christian missionaries, had to be allowed into China's exterior for the first time. The Qing rulers, overall, remained hostile to the foreign presence within China, trying to minimize it as much as their new, weaker status allowed. Within China itself, the ordinary population showed little enthusiasm for the arrival of foreigners in their midst, regardless of whether they were bringing guns or bibles with them.

The foreign presence often had unexpected results. One of the most notable was the Taiping War of 1851–64. Influenced by missionaries, a delusional failed examination candidate named Hong Xiuquan from Guangdong announced that he was the younger brother of Jesus, and that he had come to lead a Christian mission to end the rule of the Manchu 'devils' of the Qing dynasty. Recruiting in China's impoverished south, his Society of God Worshippers quickly attracted tens of thousands of followers. Hong declared that he was establishing the Taiping Tianguo (the Heavenly State of Great Peace), and his army swept through China. By the early 1860s, the Taiping was effectively a separate state within Qing territory, with its capital at Nanjing, in charge of much of China's cultural heartland. The regime was ostensibly Christian, but its interpretation demanded the recognition of Hong's semi-divine status, and Taiping rule was harsh and coercive. However, the regime did manage the remarkable feat of conquering a huge area of central China for nearly eight years. For a while, it looked as if the Taiping might bring the Qing crashing down. Certainly Karl Marx had hopes of this, and of aftershocks even further afield, writing in a New York newspaper in 1853: 'The Chinese revolution will throw the spark into the overloaded mine of the present industrial system and cause the explosion of the long-prepared general crisis.' Eventually, the retraining of local armies by loyal generals such as Zeng Guofan, as well as the stresses within the Taiping movement itself, brought the rebellion to an end, although not before countless people had died in what was perhaps the bloodiest civil war in history: contemporary accounts suggest that 100,000 people died in the final battle of Nanjing alone.

The following decades did see the Qing make efforts to reform its practices, and the 'self-strengthening' movement of the 1860s involved notable attempts to produce armaments and military technology along Western lines. Yet imperialist incursions continued, and the attempts at 'self-strengthening' were dealt a brutal blow during the Sino-Japanese War of 1894–5. Fought between China and Japan (the latter was now a fledgling imperial power in its own right) over control of Korea, it ended with the humiliating destruction of the new Qing navy, and the loss not only of Chinese influence in Korea, but also the cession of Taiwan to Japan as its first formal colony.

Most general histories, not least those written in China itself, have been highly dismissive of the last decades of Qing rule, regarding it as a period when a corrupt dynasty that refused to adapt to a new and hostile world was finally overthrown. For years, Marxist Chinese historians viewed the period as 'feudal', and argued that its overthrow set the stage for a new 'modern' era that would eventually usher in the rule of the CCP. For this reason, it was essential to portray everything in China before Mao came to power as 'feudal' or in some sense a failed modernity. However, it is now clear that significant steps towards modernity were taken in the late Qing.

One reason was that there was a powerful Asian example of how reform might be carried out: Japan. The island country across the sea would remain in Chinese minds as both dangerous menace and modern mentor for decades, just as it looms large in Chinese minds in the contemporary era. The events that had inspired and concerned the Chinese had followed the Meiji Restoration of 1868. A group of Japanese aristocrats, worried by ever-greater foreign encroachment on Japan, had overthrown the centuries-old system of the Shogun, who acted as regent for the emperor. Instead, they 'restored' the emperor to the throne under a new reign-title of 'Meiji' ('brilliant rule'), and governed in his name. These aristocrats swiftly determined that the only way to protect Japan was to embrace an all-out programme of modernization. They showed little of the ambiguity that conservatives in the Qing court had done. In quick succession, Japan replaced its culture of elite samurai warriors with a conscripted citizen army; the country was given a constitution that established it as a nation-state; and a parliamentary system was set up, although with a heavily limited male-only franchise. Modernization did not mean abandonment of Japan's past, however; the traditional folk religion of Shinto was reconstituted as State Shinto, a more formalized religion that would give spiritual sustenance to the nation. Meiji Japan also intervened heavily in the economy. The end results were clear. By the first decade of the 20th century, Japan was a growing economic and imperial power which was even able to defeat a western power, Russia, in the war of 1904–5. These headily swift changes in a country which the Chinese had always regarded as a 'little brother' gave Chinese reformers plenty of material for consideration.

One of the boldest proposals for reform, which drew heavily on the Japanese model, was the programme put forward in 1898 by reformers including the political thinker Kang Youwei (1858–1927). Kang was driven by the conviction that the previous vision of Chinese modernity, based on 'self-strengthening', had failed because it had not been comprehensive enough in its aims. Kang illustrated to the emperor the need for more thorough reform by putting forward two contrasting case studies: Japan, which had reformed successfully, and Poland, a state which had failed so comprehensively that it had disappeared from the map, carved up by powerful neighbours in 1795. The reforms were not just led from the top. Among the phenomena that emerged from that period of change were a greater participation by lower-level Chinese elites in the demand for popular rights, a new flourishing of political newspapers, and the establishment of Peking University, which remains to this day the most prestigious educational institution in China. The reformers also strongly advocated changes in the position of women. However, in September 1898, the reforms were abruptly halted, as the Dowager Empress Cixi, fearful of a coup, placed the emperor under house arrest and executed several of the leading advocates of change.

Two years later, Cixi made a decision that helped to seal the Qing's fate. In 1900, North China was rife with rumours of spirit possession and superhuman powers exercised by a mysterious group of peasant rebels known as 'Boxers'. Unlike the Taiping, the Boxers were not opposed to the dynasty. Rather, they wanted to expel the influences that they believed were destroying China from within: the foreigners and Chinese Christian converts. In summer 1900, China was convulsed by Boxer attacks on these groups. Fatefully, the dynasty declared in June that they supported the Boxers, relabelling them as 'righteous people'. Eventually, a multinational foreign army forced its way into China and defeated the uprising. The imperial powers then demanded compensation from the Qing: the execution of officials involved with the Boxers, and a sum of 450 million taels (US$333 million) to be repaid over 39 years. The Boxer Uprising marked the last time, until Mao's victory, that a Chinese government made a serious attempt to expel foreigners from China's territory. Unlike Mao half a century later, the Qing failed.

That failure, and the huge financial burden and political disgrace which it had brought upon the dynasty, led to the most single-minded attempt at modernization that the Qing had ventured: yet another reinterpretation of what modernity meant in a Chinese context. In 1902, the Xinzheng ('new governance') reforms were implemented—this was the 'new China' of the 1910 book with which the first chapter began. The reform was a remarkable and comprehensive set of changes to China's politics and society, which in many ways echoed the abortive 1898 reforms of just four years before.

This set of reforms, now half-forgotten in contemporary China, looks remarkably progressive, even set against the standards of the present day. In 1900–10, elections were proposed at the sub-provincial level, to be held in 1912–14, with the promise of an elected national assembly to come. The elections never happened because of the republican revolution of 1911, but it is possible now to look back and imagine an alternative world in which Qing China transformed into a constitutional monarchy, as indeed did happen in its south-western neighbour, Siam (Thailand), in 1932. The more immediate example to hand was Japan, and it is notable that many of the reforms of the period, such as in education, technology, and the police and military were heavily shaped by Chinese who had learned from the Japanese example.

The elections were limited (as were most such elections in the West at the time) to men (not women) of property and education, but the granting of such rights to any group showed how far society had changed in the last years of the Qing. Figures such as Zhang Jian, who set up factories in Nantong, a small city near Shanghai, were becoming typical of a new commercial middle class, and the dynasty actively encouraged the formation of bodies such as Chambers of Commerce to articulate the interests of such groups.

The most significant cultural shift in the reforms came in 1905, with the abolition of the almost thousand-year-long tradition of examinations in the Confucian classics to enter the Chinese bureaucracy. When it had first been implemented, the objective, rational standards of the entrance examination had made the system far more impressive than anything the rest of the world could offer in deciding who would govern; but by the early 20th century, the system had become inflexible, and the term 'eight-legged essay', referring to a standardized form of composition which the candidates had to write, had become synonymous with backwardness and conservatism to many reformers. In 1905, the dynasty replaced the system with alternative examinations in science and languages.

So there were significant reforms during what turned out to be the last decade of Qing rule. Nonetheless, the dynasty *did* collapse, unlike the Japanese empire of the time. Why, if the dynasty was not simply a corrupt shell, and had real potential for reform, did it do so?

First, the economic crisis of the late Qing was real, and in particular, there was a problem with agricultural productivity. Historians trace its origins to the late 18th century, meaning that the foreign invasions exacerbated, but did not create the crisis itself. In addition, the imposition of favourable tariff rates for the foreign powers meant that China's capacity to produce competitively in its domestic market or for export was hampered. Despite later arguments that the impact of imperialism actually helped China to develop, the British and French were clearly not investing in China to assist the Chinese economy, but rather to boost their own imperial projects. (Japan, which swiftly renegotiated treaty rights that were over-favourable to foreigners, saw its much smaller economy grow much faster over the same period.) In addition, the Qing made a particularly unwise choice in 1900 when it supported the Boxer Uprising.

After the failure of the Boxers, the Qing dynasty was forced to pay an immense indemnity, a financial burden that led in part to the initiation of the Xinzheng reforms. Taxation revenue continued to be unreliable and marred by massive corruption in the late Qing.

Second, ever since the Taiping, authority in China had become much more localized and militarized. The huge increase in the Chinese population during the Qing had made it ever harder for the bureaucracy to cope with administering society as a whole. Tax collection, the basis on which any society operates, had become insufficient and corrupt, with local officials adding 'surtaxes' that lined their own pockets rather than going into the state's coffers. Silver shortages also led to inflation, causing further tax rebellions.

Local elites had been instrumental in forming New Armies from the 1860s that allowed the threat from the Taiping and other rebels to be beaten back. But this moved influence away from the central government and squarely towards to the provinces. This would be a factor when the empire finally did collapse: the ground had been set for China to divide into feuding provinces led by warlords, each with his own local army, something that would have been harder to imagine in 1800.

Finally, it may have been the reforms themselves that doomed the dynasty. Empires often collapse when they try to reform, and unleash a forum for voices that are hostile to those in authority. It was in 1989, during the most liberal era of Communist rule, and not in 1969, at the height of the Cultural Revolution, that protesters filled Tian'anmen Square demanding that the leadership step down. It was under the liberal Gorbachev, not the brutal Stalin, that the USSR finally collapsed.

The abolition of the examinations in 1905, for instance, created a huge number of angry local elites. For centuries, men would spend years learning the Confucian classics in the hope that they might pass the increasingly severe level of examinations that would let them rise to local and even national status in the bureaucracy. Thousands of young (and not so young) men took these examinations and very few passed. But now, the government abolished their *raison d'être* at a stroke. From 1898 onwards, with the sudden ending of a promising series of reforms, too many of China's elites no longer trusted the Qing to reform China successfully. The Xinzheng reforms were not too little, too late, but perhaps too much, too late.

Among the figures dedicated to ending, rather than reforming, the dynasty's rule was the Cantonese revolutionary Sun Yatsen

(1866–1925). Sun and his Revolutionary League made multiple attempts to undermine Qing rule in the late 19th century, raising sponsorship and support from a wide-ranging combination of diaspora Chinese, the newly emergent middle class, and traditional secret societies. In practice, his own attempts to end Qing rule were unsuccessful, but his reputation as a patriotic figure dedicated to a modern republic gained him high prestige among many of the emerging middle-class elites in China. As it turned out, however, his stock was less high among the military leaders who would have China's fate in their hands for much of the early 20th century.

The end of the dynasty came suddenly, and had nothing directly to do with Sun Yatsen. Throughout China's south-west, popular feeling against the dynasty had been fuelled by reports that railway rights in the region were being sold off to foreigners. A local uprising in the city of Wuhan in October 1911 was discovered early, leading the rebels to take over command in the city and hastily to declare independence from the Qing dynasty. Within a space of days, then weeks, most of China's provinces did the same thing. Provincial assemblies across China declared themselves in favour of a republic, with Sun Yatsen (who was not even in China at the time) as their candidate for president. Yuan Shikai, leader of China's most powerful army, went to the Qing court to tell them that the game was up: on 12 February 1912, the last emperor, the 6-year-old Puyi, abdicated.

The crisis of the republic

In 1912, the Republic of China was declared. In 1949, Mao Zedong announced the establishment of the People's Republic of China. The 37 years of the Republic founded in 1912 were dismissed by the Communists as a time of failure and broken promises, and in general, they continue to be regarded as a dark time in China's modern history. Certainly there was much to condemn during the period: poverty, corruption, and China's weakness in the international system. Yet, as in Weimar Germany, another period of political turmoil, the chaos actually cleared space for new ideas and a powerful cultural renaissance to begin. In terms of freedom of speech and cultural production, the Republic was a much richer time than any subsequent era in Chinese history. Even in political and economic terms (see also Chapter 5), the period demands serious reassessment. Yet one cannot deny that the high hopes that the revolutionaries had for their Republic were swiftly dashed.

Sun Yatsen had returned to China from his trip abroad when the 1911 Revolution broke out, and briefly served as president, before having to make way for the militarist leader Yuan Shikai. In 1912, China held its first general election, and it was Sun's newly established Nationalist (in Chinese, *Guomindang* or *Kuomintang*) Party which emerged as the largest party. Parliamentary democracy did not last long. The Nationalists' prime-minister-in-waiting, Song Jiaoren, was assassinated at Shanghai railway station, and shortly afterwards, the Nationalist Party itself was outlawed by Yuan Shikai. Sun had to flee into exile in Japan, not to return until 1917, after Yuan had died. However, the death of Yuan and the lack of parliamentary representation meant that there was no unifying leader for China, and the country split into rival regions ruled by militarist leaders ('warlords') from 1916 onwards. Whoever controlled Beijing was recognized by the international community as the official government of China, and was also entitled to national tax revenues from the Maritime Customs Service, the unusually structured agency that brought in income for the national government (it was an institution of the Chinese state, but largely staffed and run at senior levels by westerners). This meant that control of the capital was a financial as well as symbolic prize. Yet governments in Beijing in reality often controlled only parts of northern or eastern China and had no real claim to control over the rest of the country.

The most notable challenge to the idea that the new Republic of China was an independent, sovereign, and modern state was the reality that foreigners still had right of veto or control over much of China's domestic and international policy. The global situation following World War I had slowed down formal European imperial expansion, but Britain, France, the US, and the other Western powers showed little desire to lose those rights, such as extraterritoriality and tariff control, which they already had. Japan, unlike the Western powers, sought further expansion: in 1915, while Europe was distracted by war, the Tokyo government made the notorious 'Twenty-One Demands' to the warlord government of the time, demanding and obtaining exclusive political and trading rights in large parts of China. To many Chinese, the weakness and venality of the Republic seemed a mockery of the whole project of modernizing China.

The city of Shanghai became the focal point for the contradictions of Chinese modernity. By the early 20th century, Shanghai was a wonder not just of China, but of the world, with skyscrapers, neon lights, women (and men) in outrageously new fashions, and a vibrant, commercially minded, take-no-prisoners atmosphere that made it famous—or notorious—as the 'pearl' or the 'whore' of the Orient. Its central areas, the International Settlement and French Concession, were out of Chinese governmental control, being run respectively by a foreign ratepayers' council (the Shanghai Municipal Council) and a French governor. This caused a conflict of feelings among many Chinese, from elite to grassroots level. The racism that came with imperialism could be seen every day, from the kicks and beatings given by boorish European travellers to rickshaw-drivers all the way to the race bar that prevented rich, bourgeois Chinese from membership in European clubs or entry to certain public parks in the city. Yet the glamour of modernity was undeniable too. Millions of migrant workers streamed from the countryside to Shanghai to make a new life. More elite Chinese came to the city to encounter French fashion, British architecture, and American movies, and to enjoy the relative freedom of publication and speech that the city seemed to afford. It is no surprise that Shanghai in the pre-war period had more

31

millionaires than anywhere else in China, yet also hosted the first congress of the Chinese Communist Party.

Overall, however, Chinese politics seemed to many to be collapsing into warlordism and regionalism by the 1920s. Yet the political situation was not all bleak, nor were the warlord armies simply the weak and ineffective soldiers they were often mocked as being by Western observers. Some militarist leaders lived up to the popular image, such as Zhang Zuolin (1875–1928) of Manchuria, who was an illiterate though shrewd and calculating former bandit who in 1923 spent 76 per cent of his region's tax revenues on warfare to increase his territory, and only 3 per cent on education. Others, however, were interested in construction as well as warfare. Yan Xishan (1883–1960) of Shanxi became well known as one of the most progressive warlords, actively promoting an anti-foot-binding campaign in his province, and Zhang Zuolin's son, Zhang Xueliang (1900–2001), oversaw a significant increase in transport, education, and commercial infrastructure in Manchuria, the north-eastern provinces of China.

Nor, despite their reputation for corruption, were the militarist governments capable only of considering their own immediate interests. A case in point is China's entry into World War I. The militarist group that held power in Beijing in 1917 had calculated that the Great War in Europe would end with the defeat of Germany, one of the great European empires which held colonial rights on Chinese territory. Prime Minister Duan Qirui argued that it would help China's cause to support the war on the Allied side. Duan's government offered combat troops, which the French were willing to accept, as the number of French casualties climbed higher by the month. The British were less keen: they had lost proportionately fewer men and were more sensitive about racial hierarchy. In the end, 96,000 served on the Western Front not as soldiers but digging trenches and doing hard manual labour. Around 2,000 of them died in Europe. The Chinese government, despite its problems and its lack of popular mandate, had made a momentous decision: China could—and should—become involved in international conflicts.

It was involvement in World War I, however, that led to one of the most important events in China's modern history: the student demonstrations on 4 May 1919 (see Box 1).

(see Box 1)

Box 1 May Fourth Movement

The news was bad. On 30 April, it became clear that the Paris Peace Conference was not going to return to Chinese sovereignty the parts of Shandong province that had been colonized by the now-defeated Germans; instead, the Allied Powers would hand them to Japan. Just five days later, on 4 May 1919, some 3,000 students gathered in central Beijing, in front of the Gate of Heavenly Peace, and then marched to the house of a Chinese government minister closely associated with Japan. Once there, they broke in and destroyed the house, and beat a visiting official so badly that his body was covered in marks that 'looked like fish-scales' on his skin.

This event, over in a few hours, became a legend. Even now, any educated Chinese will understand what is meant by 'May Fourth'—no year necessary. For the student demonstration came to symbolize a much wider shift in Chinese society and politics. The new Republic had been declared less than eight years earlier, yet already the country seemed to be falling apart because of imperialist pressure from outside and warlord government within China that had destroyed its fragile parliamentary democracy. The May Fourth Movement, as it became known, was associated closely with the 'New Culture' which intellectuals and radical thinkers proposed for China, to be underpinned by the twin panaceas of 'Mr Science' and 'Mr Democracy'. In literature, a 'May Fourth' generation of authors wrote fiercely

(continued)

The old order and the new

Box 1 Continued

anti-Confucian works, condemning the old culture that they felt had brought China to its current crisis, and explored new issues of sexuality and individual selfhood (see Chapter 6). In politics, young Chinese turned in desperation to new political solutions: among them, the Nationalist Party of Sun Yatsen, and for the more daring, the fledgling Chinese Communist Party, founded in 1921. Many of its founder members, including the young Mao Zedong, had been associated with the intellectual ferment at Peking University, whose students had been prominent in the May Fourth demonstrations. In the decades since, the CCP has become the world's largest governing party, and has long since become the establishment in Chinese politics. Yet it still regularly attributes its origins to the rebellious students who marched on 4 May 1919.

Double-dealing by the Western Allies and Chinese politicians who had made secret deals with Japan led to an unwelcome discovery for the Chinese diplomats at the Paris Peace Conference: the former German colonies on Chinese soil would not be returned to Chinese sovereignty, but instead would be handed over to the Japanese, who had also entered the war on the Allied side in 1917. Patriotic demonstrations by students in Beijing on 'May Fourth' became symbolic of a wider feeling of national outrage that China was being weakened internally by its unstable, militarist governments, and externally by the continuing presence of foreign imperialism.

The outrage symbolized by the May Fourth demonstrations gave rise to a whole range of new thinking collectively termed the 'New Culture' movement, which stretched from around 1915 to the late 1920s. In China's cities, literary figures such as Lu Xun and Ding Ling wrote fiction that was designed to alert China to its state of crisis (see Chapter 6). Political thinkers turned to a variety of -isms, such as liberalism, socialism, and anarchism, and also sought inspiration in a variety of foreign examples, including nationalist figures such as Washington and Kossuth, but also non-European figures such as Gandhi and Atatürk. The shooting of striking factory workers in Shanghai by foreign-controlled police on 30 May 1925 (the 'May Thirtieth Incident') inflamed nationalist passions yet further, giving hope to Sun Yatsen's Nationalist party, now regrouping under the protection of a friendly warlord in Canton.

The 'ism' which emerged at this time and would later become dominant was, of course, communism. The Chinese Communist Party (CCP), later the engineer of the world's largest peasant revolution, started with tiny, urban roots. It was founded in the intellectual turmoil of the May Fourth Movement, and many of its founder figures were associated with Peking University, such as Chen Duxiu (dean of humanities), Li Dazhao (head librarian), and a young Mao Zedong (a mere library assistant). In its earliest days, the party was more like a discussion group of like-minded intellectuals with few members, although it was still politically dangerous to take part in its activities. Few of its members yet had a strongly theoretical view of Marxism. The process that would help turn the CCP into a machine to rule China would be stimulated by the intervention of Soviet assistance. Before that happened, the CCP would find itself in alliance with the Nationalist leader, Sun Yatsen.

The Northern Expedition

Sun himself did not return to China as a national leader in 1917. Instead, he was forced to rely on the support of a militarist leader in Guangdong province, Chen Jiongming, who was sympathetic to Sun's ultimate aim of reuniting China, and allowed him a base in Canton. The other key source of support for Sun was international. Sun had tried in vain to gain Western and Japanese assistance, but in 1923, he was able to gain formal support from the world's newest and most radical state: Soviet Russia (later the USSR). The Soviets did not think that the fledgling CCP, which they advised from its foundation in 1921, had any realistic prospect of seizing power in the near future. Therefore, they ordered the party to ally itself with the much larger 'bourgeois' party, the Nationalists. At the same time, their alliance was attractive to Sun: the Soviets would provide political training, military assistance, and finance. From the Canton base, the Nationalists and CCP trained together from 1923 in preparation for their mission to reunite China.

Sun himself died of cancer in 1925. The succession battle in the party coincided with the sudden rise in anti-foreign feeling that came with the May Thirtieth demonstrations and boycotts. Under Soviet advice, the Nationalists and CCP prepared for their big push north in 1926, the 'Northern Expedition' that was supposed finally to free China from splits and exploitation. In 1926–7, the Soviet-trained National Revolutionary Army made its way slowly north, fighting, bribing, or persuading its opponents into accepting Nationalist control. The most powerful military figure turned out to be an officer from Zhejiang named Chiang Kaishek (1887–1975). Trained in Moscow, Chiang moved steadily forward and finally captured the great prize, Shanghai, in March 1927 (see Figure 3).

However, there was a horrific surprise in store for his Communist allies. Chiang's opportunity to observe the Soviet advisers close-up had not impressed him. He was convinced (not without reason) that their intention was to take power in alliance with the Nationalists and then thrust the latter out of the way to seize control, Bolshevik-style, on their own. Instead, Chiang struck first. Using local thugs and soldiers, Chiang organized a lightning strike that rounded up Communist Party activists and union leaders in Shanghai, and killed thousands of them with immense brutality. Chiang's actions were part of a wider tapestry of violent

3. **Chiang Kaishek was China's leader from 1928 to 1949, and ruled over Taiwan until his death in 1975. He is shown here decorating soldiers during the war against Japan (1937–45).**

conflict that rocked China throughout this period: in the south, where the CCP had the upper hand, there were massacres of Nationalist supporters.

The Nationalist government of Chiang Kaishek was born in blood. Yet it deserves a more objective assessment than it has had until recently. In many ways, as Chapter 3 suggests, the Nationalists under Chiang and the Communists, eventually led by Mao, had much in common. Both parties saw themselves as revolutionary, and both would swiftly conclude that their revolutions had come grinding to a halt. The slogan of the Nationalist party—'the revolution is not yet complete'—could have been uttered with equal conviction by Mao. It was their similarity of intention, in part, that made their rivalry so deadly.

Chapter 3
Making China modern

The story of politics in 20th-century China has usually been told as a narrative of conflict: in particular, the conflict between the Nationalists under Chiang Kaishek and the Communists under Mao Zedong. Certainly, the clash between these two parties shaped Chinese life for decades. Chiang's China was marked by the rhetoric of *jiaofei*—bandit extermination—a reference to the elimination of the Communists, whom he refused to dignify even with the term 'party' until forced into an alliance with them during the anti-Japanese war. In Mao's China, it was Chiang who was the bogeyman: the corrupt feudal warlord whose family had exploited China for all it was worth until the Communists had swept the country clean in 1949.

Decades after the deaths of Mao and Chiang, it is possible not only to look at those two major figures with some perspective, but also to pay more attention to the context around them. There is an alternative to regarding the early 20th century as a clash of the two Chinese giants: instead we can treat the period from the establishment of Chiang's Nationalist government in 1928 to the present day as one long modernizing project by two parties that agreed as well as disagreed. Both the Nationalists and the Communists wished to establish a strong centralized state, remove imperialist power from China, reduce rural poverty, maintain a one-party state, and create a powerful industrialized infrastructure in China. Both parties launched powerful campaigns against 'superstition', believing that 'backward' spiritual beliefs were preventing China from reaching modernity. The major ideological difference was that the Chinese Communist Party (CCP) believed that none of these goals, especially rural reform, was possible without major class warfare. The Nationalists opposed this, in part because it was captive to forces that opposed economic redistribution. This division led to a deadly falling-out by the mid-1920s, which was resolved only by the Communist victory in 1949. Ironically, though, by the end of the century, the CCP had also abandoned class war, although only after decades of factional, often highly destructive, conflict between classes.

The Nationalists in power

Chiang Kaishek's Nationalist government came to power in 1928 through a combination of military force and popular support. Chiang's bloody tactics had been displayed at Shanghai, then at Canton, in 1927, when his henchmen turned on his former Communist allies and had thousands of them killed. This action seemed to many to be an augury of the type of government that Chiang would run: keener to use violence than persuasion. His government spoke of the 'people's rights', one of the Three Principles of Sun Yatsen's political philosophy (along with 'nationalism' and 'the people's livelihood'), but it suppressed political dissent with great ruthlessness, using arbitrary arrest and torture, the latter technique characterized by the activities of the Nationalist secret service chief, Dai Li. Nationalist governance was marked by corruption and frequent capitulation to the demands of those who had a vested interest in the old order. An indicative figure is that in 1930, China's mortality rate was the highest in the world, more than that of colonial India and 2.5 times that of the US.

Yet the Nationalist record in office also had significant strengths which have generally been overlooked. In particular, many of the aspects of contemporary China which have attracted interest today, as well as achievements for which Mao has been given credit, actually originated with the Nationalist government. Chiang's government began a major industrialization effort, greatly augmented China's transport infrastructure, and successfully renegotiated many of the 'unequal treaties' which had so blighted relations between China and the imperial powers since the Opium Wars.

Throughout its life, though, the government suffered from one crippling reality. Its status as the 'National Government' of China was internationally recognized. Yet it never really controlled more than a few (albeit very important) provinces, although Chiang's level of influence in western China grew by the mid-1930s as he allied with regional militarists to drive the Communists north. Regional militarists continued to control much of western China; the Japanese occupied Manchuria in 1931; the Communists re-established themselves in the north-west. Yet even Chiang's partial consolidation was quickly shattered in 1937. 'Free China', the Nationalist-controlled interior of the country, was confined to parts of southern and central China during the war years. The final reunification in 1945 came too late for the government to take advantage of it, and the crippled Chiang government gave way to the Communists just four years later. Chiang was permanently hobbled by having to act as head of a country which was actually significantly disunited.

Furthermore, the causes of some of the horrific realities of early 20th-century China were not fully understood at the time. It was commonplace to argue that China's problems stemmed in part from too large a population and too little food. Yet it now seems that, although famines did hit parts of China on a regular basis, there was no chronic shortage of food. Rather, appalling levels of hygiene and health care before 1949 meant that mortality rates were high even among a population which had enough to eat overall. Hygiene campaigns under the CCP after 1949 heavily reduced mortality rates even when the population was much larger than a few decades earlier. Nor was the economy as disastrous as has sometimes been implied (see also Chapter 5). It was assumed for a long time that the traditional handicrafts practised in the Chinese countryside were destroyed by imperialism and the mechanization it brought along with it. Yet we now know that some traditional handicrafts did remain strong into the 1930s. Although the spinning of yarn by hand had been made obsolete by the development of mechanized cotton mills, the weaving of high-quality handmade cotton and silk cloth was of considerable importance in China's pre-war economy.

To make a balanced assessment of the government which ruled China from 1928 to 1949, it is important not to see the Nationalists simply as their enemies perceived them, but also in the terms in which they saw themselves. Chiang believed entirely sincerely that his role was to carry out the 'unfinished revolution' of Sun Yatsen. In this vision, China would be united and the militarist divisions which had torn it apart for two decades would end. China would have a full role in the community of nations, but colonialism on its soil would no longer be permitted. The Nationalist vision for China also saw it as an industrialized state: indeed, it was Sun Yatsen who first suggested damming the Yangtze to provide electric power, a goal that would eventually be fulfilled some eight decades later when the Three Gorges Dam opened in 2003.

The Nationalist vision of China was indubitably modern: it wished to create a self-aware citizenry which would live in a rational, scientific way. For this reason, the Nationalists spent significant energy combating 'superstition': traditional folk customs and religious practices which they felt were out of step with a modern China. Despite his desire to restore some of the supposed traditional values of China that had been undermined by modernity, Chiang Kaishek also spoke of 'science' as a force that could transform the country into a powerful, independent state.

Chiang's government has often been criticized as being in the pay of China's emergent capitalist class. In fact, Shanghai's capitalists had a wary relationship with Chiang, who was keen to raise revenue from them, and was not above using extortion to do so. More fundamentally, the Nationalists began a project of state-sponsored industrialization and development, much of it in cooperation with the League of Nations, which provided significant technical assistance, the greatest success of which was probably the flood control measures undertaken after the disastrous Yangtze river floods of 1931. The measures were extensive enough that the next big floods, in 1935, caused relatively little loss of life. In its first two years, the Nationalist government also managed to double the length of motor highway in China and increase the number of students studying engineering. State planning has generally been associated with the socialist bloc during the Cold War, but in fact it was widely advocated in democracies and dictatorships alike during the interwar period (Roosevelt's New Deal being a notable example), particularly as the Great Depression encouraged protectionism and turned governments against the idea of letting the unfettered market decide. State planning was at the heart of the Nationalists' desire to change China through modernization.

In 1934, the Nationalists had succeeded in forcing the Communists onto the Long March away from Jiangxi province, a remote, rural part of central China. It was here that Chiang launched his own attempt at an ideological counter-argument to communism: the New Life Movement. This was supposed to be a complete spiritual renewal of the nation, through a modernized version of traditional Confucian values, such as propriety, righteousness, and loyalty. In terms of personal behaviour, the New Life Movement demanded that the renewed citizens of the nation must wear frugal but clean clothes, consume products made in China rather than seek luxurious foreign goods, and behave in a hygienic and ordered way (no wandering out randomly into the roads or urinating in public places). The

43

Movement's aims, though, were not traditional but modern: it sought to be a mass movement that would produce a militarized, industrialized and more culturally aware China. Madame Chiang Kaishek said of it in 1935: 'mere accumulation of wealth is not sufficient to enable China to resume her position as a great nation. There must also be a revival of the spirit, since spiritual values transcend mere riches.' The aim of the movement was to create a citizenry that was self-aware, politically conscious, and committed to the nation. The policy itself was derived from a variety of sources, including Confucianism, muscular Christianity, and Social Darwinism. Despite its anti-communism, it shared many values and assumptions with the CCP, with its stress on frugality and collective values. Yet it never had much success. While China suffered from a massive agricultural and fiscal crisis, prescriptions about clothes and orderly behaviour did not have much popular traction.

Why did the grand plans of the Nationalists fail, as those of the Qing had failed? Technological modernity was all very well, but much of its impact was confined to the cities, and did little to change life on the ground in the countryside, where over 80 per cent of China's people lived. (There were exceptions to this, and the role of the railway in shrinking distances within China in the early 20th century should be noted.) The Nationalists did undertake some rural reform, including the establishment of rural cooperatives in several provinces of China, although their effects were small, with only around three-quarters of a million farmers in cooperatives by 1935. The party also quickly became entwined with the interests of local groups who had little interest in reform, and made unwise decisions that weakened their rule. The biggest obstacle to the state's aims was its inability to gather tax revenue. Unable to create a strong central revenue agency, it was reliant on other agencies, including the Maritime Customs, a hybrid organization, part of the Chinese government, but headed by a foreigner (usually a Briton), to collect revenue on goods in transit for the Chinese government, and 'tax farming'. The latter meant the devolution of tax collection to local elites, who were then free to extort and bully payments from the wider population. By devolving its tax-collection enterprise in this way, the Nationalists solved a short-term problem only by an immensely destructive long-term corrosion of public trust in the state's ability to operate honestly and efficiently.

But one factor above all affected the ability of the Nationalists to establish any kind of stable and effective government: the Sino-Japanese War of 1937–45, known in China as the War of Resistance against Japan. Less than a decade after its establishment, Chiang's government was plunged into total conflict.

The eight years of war devastated China. Exact mortality figures have never been worked out, but a minimum number appears to be around 14 million (with some estimates of dead and wounded as high as 35 million). Some 80 million or more Chinese became refugees. The slowly expanding technological and industrial infrastructure of the previous decade was destroyed (some 52 per cent of the total in Shanghai, and some 80 per cent in the abandoned capital of Nanjing). The government had to operate in exile from the far south-western hinterland of China as its area of greatest strength and prosperity, China's eastern seaboard, was lost to Japanese occupation. At the same time, the Communists, who were forced into the remote north-west after the Long March of 1935–6, now found themselves able to consolidate and expand their hold on northern China. Chiang's project of consolidation and unification was thrown into reverse gear. Wartime conditions encouraged corruption, black marketeering, and hyperinflation.

The Japanese invasion of China which began in 1937 was merciless (see Figure 4). Among the many atrocities committed against the population, the most notorious is the Nanjing Massacre ('the Rape of Nanking') that took place in the weeks

4. The Sino-Japanese War of 1937–45 tore China apart. Here refugees stream through the wartime capital of Chongqing after a heavy air-raid.

between December 1937 and January 1938. The Nationalists had had to abandon their capital and the city was left defenceless when Japanese troops arrived at the gates. Out-of-control soldiers, unrestrained by their commanders, indulged in weeks of mass killings, rapes, and destruction of property that resulted in tens, and quite possibly hundreds of thousands of deaths. Yet this was just one of a series of war crimes committed by the Japanese Army during its conquest of eastern China. Nonetheless, the Japanese succeeded in gaining cooperation from some Chinese, many of whom felt that defiance of the Japanese would bring down yet further horrors upon them. This was the justification used by one of Chiang's long-time Nationalist rivals, Wang Jingwei, who defected to the Japanese in 1938 and was made president of a 'reorganized' Nationalist government at Nanjing in 1940. Wang's government portrayed itself as the real heir to Sun Yatsen, and denounced Chiang as a traitor who had allied with the hated Communists, but his regime had no status independent of collaboration with Japan, and never achieved mass support. Wang himself died of cancer in 1944.

Although Japan was defeated in 1945, post-war China was still in a state of shock. Unlike the pre-war period, food was in genuinely short supply in 1945, as much of China's agricultural heartland had been destroyed, and famine relief supplies from abroad were able to deal with only a small part of the problem. Scarcity caused inflation, to which the government responded by printing money, in part to pay for its huge military commitments: a foolish move, but alternative policies were hard to think of. A large sack of rice increased in price from 6.7 million yuan in June 1948 to 63 million yuan just two months later. By the time China was plunged into yet another war—the Civil War between the Communists and the Nationalists (1946–9)—the huge disillusionment with Nationalist rule meant that many who were not Marxists by inclination welcomed a Communist victory simply because they felt that the Nationalists had no credibility left.

During all this time, the Communists (CCP) had not stood still. After Chiang had turned on them in the cities, most of what remained of the party stole into the countryside, away from Chiang's area of control. A major centre of activity was the base area in Jiangxi province, an impoverished part of central China. It was here, between 1931 and 1935, that the party began to try out systems of government that would eventually bring them to power, and shape the People's Republic. After their experience with Chiang, the party felt it essential to train a Red Army of their own. They also experimented with land redistribution and representative government, although they were wary of alienating local elites while the party was still vulnerable, and therefore did not push policies that might lead to local leaders turning against them. It was during this period that Mao first began to gain power: while he had been one of the earliest members of the CCP, it was the period in the rural areas that allowed him to come to the fore. However, it was also Mao's influence that contributed to increased intra-party violence and more radical attacks on local landowners in the mid-1930s. In addition, by that time, Chiang's previously ineffective, if uncompromisingly named 'Extermination Campaigns' were beginning to make the CCP's position in Jiangxi untenable.

In 1934, the party began the action that remains a legend to the present day: the Long March. Travelling over 4,000 miles, some 8,000 of the more than 80,000 Communists who had set out finally arrived, exhausted, in Shaanxi province in the north-west. They were safe, but they were also on the run. It seemed possible that within a matter of months, Chiang would once again attack (see Figure 5).

5. The Long March of 1934–5 helped Mao rise to paramount leadership of the Chinese Communist Party. This image is taken from a tapestry depicting scenes from the March.

The approach of war saved the CCP. There was growing public discontent at Chiang's unwillingness to fight Japan. In fact, this perception was somewhat unfair. The Nationalists had undertaken retraining of key regiments in the army under German advice, and also started to plan for a wartime economy from 1931, spurred on by the invasion of Manchuria. However, Chiang knew that it would take a long time for China to be capable of resisting the well-trained armies, superb technology, and colonial resources of Japan, and preferred an unglamorous but practical solution of diplomatic appeasement. By 1936, however, this was no longer feasible. Events came to a head in December 1936, when the militarist leader of Manchuria (Zhang Xueliang) and the CCP managed to kidnap Chiang. After negotiations, Chiang confirmed a United Front, which had already been under discussion, in which the Nationalists and Communists would sink their differences and ally against Japan.

This development gave the CCP valuable breathing space. They had several areas under their control in wartime China, but the single most famous was the ShaanGanNing base area with its capital in the small town of Yan'an. However, it was the presence of one figure in particular that made 'Yan'an' an iconic term: Mao Zedong. His credibility as a political figure had been greatly boosted because his advocacy of revolution using the peasantry seemed wiser than the now-discredited, Soviet-backed policy of urban revolution. Yan'an was not easy to get to, its isolation being an advantage in protecting the area from the Nationalists and the Japanese. Mao took advantage of this, using the period to implement a variety of policies that would eventually influence his rule over China. These included attempts to create a self-sufficient economy, tax and land reforms to relieve the poverty of the rural population, and fuller representation for the local population in government. At the same time, Mao reshaped the party in his own image. The Party was purified through repeated 'Rectification' (*zhengfeng*) campaigns, which sought to impose an ideological purity on party members based in Mao's own thought, rather than encouraging dissent.

The American journalist Edgar Snow, who sympathized deeply with the Party, travelled to Yan'an and met Mao. In his classic account of that journey, *Red Star over China*, he noted in adoring terms of Mao:

> He had the simplicity and naturalness of the Chinese peasant, with a lively sense of humor and a love of rustic laughter…he combined curious qualities of naivete [*sic*] with incisive wit and worldly sophistication.

By the end of the war with Japan, the Communist areas had expanded massively, with some 900,000 troops in the Red Army, and Party membership at a new high of 1.2 million. For much of the 1930s and into the war years, it was clear that there were two major ideological poles for nationalists who opposed the Japanese presence: Chiang's government at Nanjing (then Chongqing, during the war), and Mao's at Yan'an. Both of these offered a powerful vision of modernity, in both cases at odds with the somewhat ramshackle reality of their areas of control. But greater power for one pole sucked it away from the other: in the mid-1930s, the Nationalist pole seemed to be gaining strength, but by the mid-1940s, war and its effects had reversed the balance.

Above all, the war with Japan had helped the Communists come back from the brink of disaster, and destroyed the tentative modernization that the Nationalists had undertaken. In the aftermath of Japan's defeat, both sides scrambled for power (see Box 2). Unable to reach a compromise, the Nationalists and Communists plunged into civil war in 1946, a war ended only by victory for the CCP in 1949. Chiang fled to Taiwan, and in Beijing, now restored to its status as the capital, Mao declared the establishment of the People's Republic of China (PRC).

Box 2 Victory in Chongqing

In the centre of the city of Chongqing you will find a tall piece of statuary, now rather dwarfed by the glass and steel skyscrapers surrounding it. This is the Liberation Monument. There are many references to 'liberation' (*jiefang*) in China today, but almost all these references are to 1949, the year that the Communists finally gained control of the Chinese mainland. But in Chongqing, the Liberation Monument originally commemorated another event: the victory over Japan in 1945. It was this moment that marked the turning point on the question: What sort of country would post-war China be?

It is worth remembering the assumptions of most Chinese and of the world at large on that day. It was not Mao but Chiang Kaishek, as China's leader through eight years of war, who received the world's accolades and held victory parades through his wartime capital at Chongqing. The assumption of Chiang, and the world, was that his government would now take up the state-building project that he had begun in the 1920s.

The West certainly assumed that Chiang would be in power for years. President Roosevelt had characterized China as one of the Four Policemen (along with the US, USSR, and Britain) that would monitor the post-war world. Prime Minister Churchill had far less enthusiasm for treating Chiang as a serious global figure, but he did not see him losing power. Certainly, China's determination to last out through the war had reaped rewards on the global stage. In 1943, all of the 'unequal treaties' that had plagued Chinese–Western relations for a century had finally been ended. In 1946, China would become one of the five permanent members of the Security Council of the new United Nations. Huge international aid efforts were made to relieve the desperate poverty in China's cities and countryside. The

(continued)

Box 2 Continued

assumption on all sides, including Stalin's USSR, was that China would remain in the Western bloc in the emergent Cold War.

Within five years, it all went wrong for Chiang. Why? Most histories have laid the blame squarely at the feet of Chiang and the Nationalists, and they must surely take a large part of the responsibility. Large swathes of the population were branded as collaborators, with little understanding shown for the dilemmas that had faced those who had to decide whether to flee or stay in the face of Japanese invasion. Many of the officials installed in the restored Nationalist government were corrupt and inefficient. Within months, the goodwill of victory was being squandered.

Was the Communist victory a military one or a social one? The two aspects are not easily separated. The Chinese Communist Party (CCP) had little chance to implement serious land reform before 1949 in most areas of China: during the Civil War, it simply did not have that level of control over the population. Without the brilliant military tactics of Lin Biao and the other PLA (People's Liberation Army) generals, social reform on its own would not have been enough to bring the CCP victory. On the other hand, it was clear by 1948 that while Chiang's armies had the support of the US and superiority in terms of finance and troop numbers, their morale was slipping away. The goodwill from victory over Japan had not been transferred to the CCP, but it was no longer with the Nationalists. Their chance for unity had been squandered.

By the time Mao declared the establishment of the People's Republic of China (PRC) in 1949, China's path was not in doubt. The year 1949 was not in itself the turning point, but the result of choices made in 1945: China was going to have a communist government allied to the USSR and largely closed to the West.

Mao in power

Mao's China was very different from Chiang's in a variety of ways. Perhaps, overall, the most powerful change was encapsulated in the slogan 'Politics in command', which was used during the Great Leap Forward campaign of 1958–62. Chiang had been concerned to create a politically aware citizenry through campaigns such as the New Life Movement, but these had failed to penetrate very successfully. Mao's China had much greater control over its population, and did not hesitate to use it. Its politics was essentially modern, in that it demanded mass participation in which the citizenry, the 'people', saw itself as part of a state project based on a shared class and national identity. The success of Mao's military and political tactics also meant that the country was, for the first time since the 19th century, united under a strong central government.

Mao's China is often characterized as isolated. It is certainly true that most Western influence was removed from the country; the large numbers of businessmen, missionaries, and educators—many of whom had spent their lives in China—were almost all expelled by 1952. However, China was now exposed to a new sort of foreign influence: the new dominance of the Soviet Union in the still-emerging Cold War. The 1950s marked the high point of Soviet influence on Chinese politics and culture: Soviet diplomats, technical missions, economists, and writers all played their part in shaping the new communist China. Yet the decade also saw rising tension between the Chinese and the Soviets, fuelled in part by Khrushchev's condemnation of Stalin (which Mao took, in part, as a criticism of his own cult of personality), as well as cultural misunderstandings on both sides. The differences between the two sides came to a head in 1960 with the withdrawal of Soviet technical assistance from China, and Sino-Soviet relations remained frosty until the 1980s.

The People's Republic used its state power in one way that was very different from the aims of Chiang's China: the pursuit of class warfare. Mao had finally risen to undisputed leadership of the CCP during the war period, and his writings made it clear that when he finally gained power, moderation and restraint were to be shunned. 'Revolution', he famously claimed in 1927 in his *Report on an Investigation of the Peasant Movement in Hunan*, 'is not a dinner party... it cannot be so refined, so leisurely and gentle, so temperate, kind, courteous, restrained and magnanimous. A revolution is an insurrection, an act of violence by which one class overthrows another.' Certainly, the years of war and the failure of the Nationalists genuinely to reform social relations in rural China meant that there was a widespread constituency in favour of violent change, driven by a conviction that previous attempts to change the structure of rural society had failed.

The initial period of 'land reform' in China in 1949–50 saw some 40 per cent of the land redistributed, and some 60 per cent of the population benefiting from the change. Perhaps a million people who were condemned as 'landlords' were persecuted and killed (see Figure 6). Yet this violence was not random. Official campaigns were instigated that oversaw and encouraged terror. The joy of 'liberation' was real for many Chinese at the time; but the early 1950s were not a golden age when China was truly at peace.

The urgency came from the continuing desire of the CCP to modernize China. The PRC had to deal with the reality of its situation in the early Cold War: the US refused to recognize the government in Beijing, and although other Western countries did open up diplomatic relations, the country was economically relatively isolated, even though it now had a formal cooperative relationship with the USSR. When relations with the Soviets also started becoming chillier in the mid-1950s, as Mao expressed his anger at the Khrushchev thaw, the CCP leaders' thoughts turned to self-sufficiency as an alternative. Mao proposed the policy

6. Land redistribution in the early 1950s was a time of joy for many peasants, but it also led to a deadly terror campaign against those judged to be 'landlords'. The landlord in this picture, taken in 1953, owned around two-thirds of an acre.

known as the Great Leap Forward. This was a highly ambitious plan to use the power of socialist economics to increase Chinese production of steel, coal, and electricity. Agriculture was to reach an ever-higher level of collectivization, with individual plots (the basis of the popular land reforms of Mao's first years in power) being subsumed into large collective farms. Family structures were broken up as communal dining halls were established: people were urged to eat their fill, as the new agricultural methods would ensure plenty for all, year after year. The Minister of Agriculture, Tan Zhenlin, declared:

After all, what does Communism mean?...First, taking good food and not merely eating one's fill. At each meal one enjoys a meat diet, eating chicken, pork, fish or eggs...delicacies like monkey brains, swallows' nests, and white fungi are served to each according to his needs...

The plan was fuelled by a strong belief that political will combined with scientific Marxism would produce an economic miracle of which capitalism simply was not capable. Yet its goal was unquestionably one of modernization through industrial technology. The stated goal of the Leap was to overtake Britain in 15 years, and by this, it was Britain's industrial capability, not its wheat fields or cattle, that was meant. The Leap has been interpreted as an example of how Mao's dominance over the CCP ended the possibilities of real debate within the leadership: when Peng Dehuai, the defence minister, tried to point out the hardships caused by the Leap at the 1959 Party conference at Lushan, he was abruptly dismissed from his post. Nonetheless, it should be noted that Mao was not alone in advocating the start of the Leap. Figures such as Chen Yun, the chief economic planner in the Politburo, were also highly supportive.

The Leap engendered great enthusiasm around the country, with Chinese in rural and urban areas alike taking part in mass campaigns that were not just economic but cultural and artistic as well. The breakdown of traditional family structures during the Leap helped to redefine women's roles, stressing their status as workers of equal standing to men (see Figure 7).

Still, the Great Leap Forward was a monumental failure. It can hardly be defined as anything else, as its methods caused a massive famine whose effects were dismissed by Mao, and caused some 20 million or more deaths. Its modernizing aims were dashed in the face of reality. Yet the return to a more pragmatic economic model in agriculture and industry when the Leap ended in 1962 did not dampen Mao's enthusiasm for revolutionary renewal as well as ideological success. By this time, the alliance with the Soviet Union had broken up in acrimony, and Soviet criticisms of Mao's radicalism spurred him on rather than restraining his actions.

This led to the last and most bizarre of the campaigns that marked Mao's China: the Cultural Revolution of 1966–76. In fact, the part

7. Chairman Mao with representatives of China's younger generation in his birthplace of Shaoshan during the Great Leap Forward in 1959.

of the Cultural Revolution that has remained in popular memory—teenage Red Guards persecuting their teachers, massive crowds of youth in Tian'anmen Square hoping for a sight of Chairman Mao—date from the early period, 1966 to 1969. But the whole decade marked the final rallying call for a particular type of self-contradictory modernity which Mao hoped to instil: an industrialized state which valued peasant labour and was free of the bourgeois influence of the city.

Mao had become increasingly concerned that post-Leap China was slipping into 'economism'—a complacent satisfaction with rising standards of living that would blunt people's revolutionary fervour. In addition, in the mid-1960s, the first generation was coming of age that had never known any life except that under the CCP, and therefore had no personal experience of the poverty and warfare that had afflicted their parents' generation. For these reasons, Mao decided that a massive campaign of ideological renewal must be launched in which he would attack his own party.

Mao was still the dominant figure in the CCP, and used his prestige to undermine his own colleagues. In summer 1966, prominent posters in large, handwritten characters appeared at prominent sites including Peking University, demanding that figures such as Liu Shaoqi (president of the PRC) and Deng Xiaoping (senior Politburo member) must be condemned as 'takers of the capitalist road'. The outside world looked on, uncomprehending, as top leaders suddenly disappeared from sight to be replaced by little-known figures such as Mao's wife Jiang Qing and her associates, later to be dubbed 'the Gang of Four'. Meanwhile, an all-pervasive cult of Mao's personality took over. A million youths at a time, known as 'Red Guards', would flock to hear Mao in Tian'anmen Square. Posters and pictures of Mao were everywhere; some 2.2 billion Mao badges had been cast by 1969. Personal devotion to Mao was now essential. For example, in June 1966, Red Guards of the high school attached to Qinghua University declared in an oath of loyalty:

We are Chairman Mao's Red Guard, and Chairman Mao is our highest leader.... We have unlimited trust in the people! We have the deepest hatred for our enemies! In life, we struggle for the party! In death, we give ourselves up for the benefit of the people! ... With our blood and our lives, we swear to defend Chairman Mao! Chairman Mao, we have unlimited faith in you!

The Cultural Revolution was not just a power grab within the leadership. The rhetoric that flowed from the Cultural Revolution showed that not only was this a movement of great ideological conviction, but one that, despite its seeming irrationality, reflected a particular type of modernity very strongly. 'The Great Proletarian Cultural Revolution now unfolding is a great revolution that touches people to their very souls', began the CCP Central Committee's 'Decision Concerning the Great Proletarian Cultural Revolution' (8 August 1966). The Red Guards, Chinese youth who were encouraged to rise up against their elders, embraced the call to revolution: 'Revolutionary dialectics tells us that the newborn forces are invincible, that they inevitably grow and develop in struggle, and in the end defeat the decaying forces. Therefore, we shall certainly sing the praises of the new, eulogize it, beat the drums to encourage it, bang the gongs to clear a way for it, and raise our hands high in welcome.' Just as Stalin had characterized the job of Party-sponsored artists in the USSR to be 'the engineers of human souls', so the Cultural Revolution was meant to provide a retooling of Chinese society to become a renewed, self-aware citizenry finally free from the shackles of the past. Like their revolutionary predecessors in the USSR and indeed in 18th-century France, the Red Guards were not ashamed to admit that their tactics were violent: a group of youths in Harbin in 1966 declared: 'Today we will carry out Red Terror, and tomorrow we will carry out Red Terror. As long as there are things in existence which are not in accordance with Mao Zedong Thought, we must rebel and carry out Red Terror!'

With its obsessive emphasis on violence as a supposedly desirable and transformative force, the Cultural Revolution was a highly modern movement. And while Mao initiated and supported it, it also had widespread support: it was a genuinely mass political movement which left many youths feeling as if they had had the best days of their lives. It was strongly anti-intellectual and xenophobic, condemning those such as doctors or teachers who were accused of being 'expert' rather than 'red', and casting suspicions on anyone who had connections with the outside world, whether the Western or the Soviet bloc. However, it also drew on Mao's conviction that the party had become soft and too comfortable with power, obsessed with urban concerns. Movements such as the 'barefoot doctor' programme took off during these years. Scorning the Ministry of Health as the 'Ministry of Urban Gentlemen's Health', Mao promoted a policy by which the peasants themselves were given the opportunity to train in basic medicine and provide health care in the villages. Although inadequate in many ways, the programme brought health care to parts of China which had had few such facilities even in the years after 1949.

Yet the Cultural Revolution could not last. Worried at the increasing violence in the streets, the party leadership ordered the People's Liberation Army (PLA), China's official armed forces, to send the Red Guards back home in 1969. The policies of the Cultural Revolution remained official dogma until 1976. However, from the early 1970s, profound changes began in Chinese politics. The long-standing policy of isolation was clearly not working. The period, however, saw a remarkable rapprochement between the United States and China, which had had no official relations since 1949; the former was desperate to extricate itself from the quagmire of Vietnam, the latter terrified of an attack from the now-hostile USSR and in turmoil after the sudden defection and death in an air crash of Mao's putative successor, defence minister Lin Biao. Secretive diplomatic manoeuvres led, eventually, to the official visit of US President Richard Nixon to China in 1972, which began the reopening of China to the West, although it would be more than a decade before ordinary Chinese and foreigners would be able to meet each other in any numbers within China itself. Slowly, the Cultural Revolution began to thaw, and that thaw would be accelerated by the death of its architect.

Gaige kaifang: reform and opening-up

Mao died in 1976. His successor was the little-known Hua Guofeng. Within two years, Hua had been outmanoeuvred by the greatest survivor of 20th-century Chinese politics, Deng Xiaoping. Deng had joined the CCP in 1924. He had been purged twice during the Cultural Revolution, but his connections had been strong enough to keep him safe, and after Mao's death, he was able to reach supreme leadership in the CCP with a programme startlingly different from that of the late Chairman. Deng took as his personal slogan the expression 'seek truth from facts', a term Mao had actually used in the 1930s, but which Deng adopted to indicate that he felt that truth and facts had been largely lacking from the political scene during the Cultural Revolution.

In particular, Deng recognized that the Cultural Revolution's profound anti-intellectualism and xenophobia were proving economically damaging to China. Deng took up a policy slogan originally invented by Mao's pragmatic prime minister and second-in-command, Zhou Enlai: the 'Four Modernizations'. This had been effectively an admission by moderates in the CCP leadership that the Cultural Revolution had led China away from the path of genuine modernization. Now, the party's task would be to set China on the right path in four areas: agriculture, industry, science and technology, and national defence.

To do so, many of the assumptions of the Mao era were abandoned. The first, highly symbolic, move of the 'reform era' (as the period since 1978 is known) was the breaking down, over time, of the agricultural collective farms that had been instituted under Mao. Farmers, in particular, were able to sell an ever-larger proportion of their crops on the free market, and it was stated explicitly that cash crops and small private plots of land were an essential part of the economy and should not be interfered with. Urban and rural areas were also encouraged to set up small local or household enterprises, with nearly 12 million such enterprises registered by 1985.

Economic equality was no longer the goal of government. As part of the encouragement of entrepreneurship, Deng designated four areas on China's coast as Special Economic Zones, which would be particularly attractive to foreign investors, thereby ending the preference for self-sufficiency that had marked the economy under Mao (see also Chapter 5).

Yet the doors were opened only a certain way. In December 1978, a young man named Wei Jingsheng used the new openness to demand 'the Fifth Modernization'—true democracy in China. He was quickly arrested and stayed in prison almost constantly until 1997, when he was exiled to the US. Nonetheless, the 1980s were marked by a remarkable openness, greater than under Mao. In politics, the drive for renewal became linked with a powerful desire to reopen China to the outside world. In 1979, full diplomatic relations were finally established between China and the US, and from the early 1980s, foreign tourists and students began to visit China in large numbers, just as a new generation of Chinese began to study and do business abroad.

Deng Xiaoping was indubitably the senior figure in the party, and he made it clear that he favoured economic reform at the fastest possible speed. In the 1980s, the USSR was still intact, but the West regarded that country as a stagnant and hostile giant, at least until the arrival of Mikhail Gorbachev to the leadership in 1985. In contrast, China seemed to be the communist giant that the West loved to love. Keen to grow the economy, friendly towards the US, Deng Xiaoping even visited Texas and wore a Stetson at a rodeo. At home, cowboys of a different sort also found their moment as the economy grew by leaps and bounds.

Notoriously, in 1985, the party chief of Hainan island, Lei Yu, had used valuable foreign exchange to import 79,000 foreign cars, 347,000 televisions, and 45,000 motorcycles for resale at inflated prices. (Yet Lei was rehabilitated in 1988, and his behaviour was only the most outrageous example of a widespread phenomenon of corruption.)

The signals that the economy should change and grow were unmistakable. In the political sphere, however, the signals were mixed. Deng had supported the rise to power of Hu Yaobang, a relatively liberal member of the Politburo, under whose influence there was a significant rise in the level of debate in journals and think-tanks which had previously been under the moribund influence of a monolithic party line. Deng was relaxed about a certain amount of ideological impurity coming along with the reforms: 'If you open the window,' he observed, 'some flies will get in.' Yet not all the leadership were as sanguine. Chen Yun, the archetypal central planner, and propaganda chief Hu Qiaomu were among those who were concerned by the materialism and ideological vacuousness that they perceived in reform-era China. They supported a series of campaigns, usually under the title of 'anti-spiritual pollution' and 'spiritual civilization', in which noxious influences from the capitalist world would be condemned. In the 1980s, politics tended to open up, only to be thrown into partial reverse after a couple of years. Yet the movement seemed to be inevitably towards a freer, market-oriented society.

The group which benefited most greatly from the reforms in contrast with the Cultural Revolution were 'intellectuals': a grab-bag category in the Chinese understanding which includes academics, and students, as well as more abstract thinkers. No longer were they termed the 'stinking ninth' (that is, the ninth class of undesirables in Cultural Revolution terminology). Instead, education was encouraged, as China strove to improve its science and technology infrastructure. Yet the new freedoms that intellectuals enjoyed gave them the appetite for more. After student protests in 1985–6 demanding further opening-up of the party, Hu Yaobang was forced to resign in 1987 and take responsibility for allowing social forces to get out of control. He was replaced as general secretary by Zhao Ziyang, who was perceived as politically less liberal, although just as strong an economic reformer. The complaints of the intellectuals were not just abstract, however. Most of them—in particular academics and students at universities—were on fixed state incomes, and as inflation began to run rife in newly rich reform-era China, they began to find that their income was rapidly becoming insufficient to cover their needs.

In April 1989, Hu Yaobang died. It was a long-standing tradition in China that the death of a well-respected figure could trigger demonstrations. In this case, students around China used the occasion of his death to organize protests against the continuing role of the CCP in public life. At Peking University, the breeding ground of the May Fourth demonstrations of 1919, students published journals with titles such as 'New May Fourth', and declared the need for 'science and democracy', the modernizing watchwords of 80 years earlier, to be revived. On 4 May 1989 itself, protesters in Tian'anmen Square held up signs, written in Chinese and English, reading 'Hello Mr Democracy!', a clear reference to the May Fourth duo, 'Mr Science and Mr Democracy', who had been tasked with China's salvation 70 years earlier.

In spring 1989, Tian'anmen Square in Beijing was the scene of an unprecedented demonstration. At its height, nearly a million Chinese workers and students, in a cross-class alliance rare by the late 20th century, filled the space in front of the Gate of Heavenly Peace (see Figure 8). The Party was profoundly embarrassed to have the world's media record events; they had been there for a historic occasion, the first visit of the reforming Soviet leader, Mikhail Gorbachev, but the event had turned to farce as Gorbachev was escorted via a roundabout route to avoid him

8. On 4 May 1989, exactly 70 years after the original May Fourth demonstrations in 1919, students once again ask for 'Mr Democracy' in Tian'anmen Square. A month later, tanks and soldiers would clear the Square by force, killing large numbers.

seeing the demonstrations. By June 1989, the numbers in the Square had dwindled only to thousands, but they showed no signs of moving. On the night of 3–4 June, the party acted, sending in tanks and armoured personnel carriers. The death toll has never been officially confirmed, but it seems likely to have been in the high hundreds or even more. Hundreds of people associated with the movement were arrested, imprisoned, or forced to flee to the West. It seemed to many that the hardliners had won, and that the chance for 'science and democracy' had ended.

China since 1989

In retrospect, now that Tian'anmen Square is more than two decades in the past, the surprising thing is what did *not* happen. China did not, as many feared, plunge into civil war; it did not reverse the economic reforms; it did not close itself off to the outside world. For some three years, politics did indeed go into a deep freeze. The liberal trends that had fuelled the protests of the late 1980s were now regarded as 'evil winds of bourgeois liberalism'. But in 1992, Deng, the man who had sent in the tanks, was now 88 years old. He must have known that his legacy might seem similar to that of Gorbachev, a reformer perceived, at least in Chinese eyes, to have failed. That year, he undertook what was ironically called his 'southern tour', the Chinese term *nanxun* referring to the emperor visiting his furthest domains. By visiting Shenzhen, the boomtown on the border with Hong Kong (and appearing to local news reporters riding a golf buggy in a theme park), Deng indicated that the economic policies of reform were not going to be abandoned. He had made other important choices. Jiang Zemin, the mayor of Shanghai, had effectively dissolved demonstrations in Shanghai in a way that the authorities in Beijing had not. He was groomed as the next senior leader, having been appointed general secretary of the Party in 1989. Furthermore, his home city, Shanghai, was finally given permission to attract foreign investment on a lavish scale, having been kept on a tight leash under Mao and in the initial period of reform. For a century, from the 1840s to the 1940s, Shanghai had been the motor of China's industry, commerce, and culture, an outward-looking metropolis that regarded itself as a world city rather than just Chinese. Now, it was being given permission to recreate the experience.

The post-Deng leadership has taken on something like a regular pattern. Jiang Zemin, the former mayor of Shanghai, became Party General Secretary in 1989. The Fourteenth and Fifteenth Party Congresses in 1992 and 1997 each signalled another five-year term for Jiang in that role, but despite strong rumours that he wished to stay on, the Sixteenth and Seventeenth Party Congresses in 2002 and 2007 confirmed Hu Jintao as Jiang's successor. It is possible also to detect changes in policy emphasis between the two leaders. Jiang's period in office was marked by a breakneck enthusiasm for economic development along with cautious political reform (for instance, the growth of local elections at village level, but certainly no move to democracy at higher levels). Between 2002 and 2012, Hu and his prime minister, Wen Jiabao, made more efforts to deal with the inequality and poverty in the countryside.

The year 2012 saw an important generational shift. In spring that year, China felt the biggest political shock in decades as the charismatic party secretary of Chongqing, Bo Xilai, fell from office amid lurid accusations of corruption, abuse of power, and even murder. Bo had been tipped by many (not least himself) to rise to the highest level of politics, the Standing Committee of the Politburo, and his fall and subsequent trial and conviction were seen as his punishment for attempting to bypass the existing power networks. In autumn that year, the leadership passed to a new party general secretary, Xi Jinping. In the following years, Xi swiftly centralized many powers under himself, thereby defining the next round of economic and political change in a highly personalized way. At the same time, his government sponsored a new rhetoric of a 'Chinese renaissance' along with a severe anti-corruption campaign that aimed at prominent targets ranging from a well-known television host to China's former top security chief, Zhou Yongkang. Chinese media and academia were also placed under new restrictions, with freedom to debate political change in terms of democracy or constitutional change being actively discouraged.

In the end, the economy will be the most important factor in deciding whether China's current political system will endure. Since the 1990s, China has embraced economic reform with a vengeance. Its politics does not have the liberal, almost naïve interest in the West that it did in the 1980s. One of the most influential intellectuals of the early 21st century, the philosopher Wang Hui, declared that the 'New Enlightenment' wave of the 1980s, which ultimately led to Tian'anmen Square, was unable to come to any understanding of the fact that China's problems are also the problems of the world capitalist market ... Finally, it was unable to recognize the futility of using the West as a yardstick in the critique of China.

But in many ways, China is far more influenced by globalized modernity than even in the 1980s. China has placed scientific development at the centre of its quest for growth, sending students abroad not in their hundreds, but tens of thousands, to study science and technology, just as the Nationalist government of the 1930s tried to develop an indigenous core of engineers and technicians.

The country also has a powerful international role in the early 21st century. China has sought economic and diplomatic influence in Africa and South America, taking advantage of a general suspicion in the Global South of the Western post-Cold War order. However, China's international influence has been hampered by two tendencies. In areas outside Asia, China's preference is for remaining neutral but friendly, and to commit to few outright statements of policy: yet crises in the Middle East, the Korean peninsula, and Ukraine, and the scramble for mineral resources in Africa and energy resources around the globe may make it harder to claim a status as a world power without providing action or using influence. Meanwhile, Chinese actions in its own backyard in the early 2010s, in particular territorial and maritime claims in the East and South China Seas, have made some neighbouring countries wary of growing Chinese influence in the region, particularly as its defence budget soars (it was US$131 billion in 2014). Although Chinese behaviour in Asia veers between hard and soft tactics, it is still unclear whether Beijing has worked out a real strategy to increase its influence in the region through consensus rather than coercion.

Nationalism has also become a popular rallying-cry at home. This does not necessarily mean xenophobia or anti-foreignism, although there are occasions (such as the reaction to the 1999 NATO bombing of the Chinese Embassy in Belgrade during the Kosovo War) that have led to violence against foreign targets and persons. But it is clear that China's own people consider that the country's moment has arrived, and that they must oppose attempts—whether by the West, or Japan—to prevent it taking centre stage in the region.

A modern politics?

A popular characterization of China's leaders of the 20th century, particularly Mao, is that they have sought to become new emperors in their own right. Although it is a colourful shorthand, such a comparison is misleading. It highlights the exotic element of Chinese politics and conceals the reality that many of the assumptions and models that have shaped Chinese political thinking in the present century are profoundly modern, and indeed, similar to those in the West.

Chinese politics since the late Qing has been dependent on nationalism, an idea that derived its legitimacy from the people as a body in their own right, and an idea that a strong state would be a rational arbiter of power. It was based on mass politics where there was a social contract between government and citizen. While the Confucian mode of governance did also embody a sort of social contract between emperor and subject, it did not think it seemly that the people should be empowered in their own right—a profoundly non-modern way of thought. This does not mean that the Confucian past simply disappeared with the arrival of modern politics. For instance, assumptions that came from a modernized Confucianism did remain, such as ideas of hierarchy and mutual obligation. But these ideas have become adapted through contact with the assumptions of modernity about mass participation and legitimacy derived from the people, as well as the importance of the individuated self.

There is a final irony that has been implied, but not stated, in the way in which this chapter has portrayed politics portrayed Chinese politics since the establishment of Chiang's Nationalist government in 1928, through the victory of the Communist Party in 1949 to the present era, as a passing on of the baton in a wider, consistent, politics of modernity. For the political form of China today—a one-party state that does still allow a significant amount of individual autonomy, a powerful state with a role in the international order which is partly cooperative and partly confrontational, and a highly successful semi-capitalist economy in which the state and party still play a significant, embedded role—means that the Communist Party of today has essentially created the state sought by the progressive wing of the Nationalists in the 1930s rather than the dominant, radical Communists of the 1960s. One can imagine Chiang Kaishek's ghost wandering round China today nodding in approval, while Mao's ghost follows behind him, moaning at the destruction of his vision. The intellectual assumptions behind both the Nationalists and the Communists in the last century were similar in many important ways, making this seeming paradox perfectly comprehensible if examined in a somewhat *longue durée* that extends back before 1978 or 1949.

Chapter 4
Is Chinese society modern?

Zou Taofen, one of the best-known journalists in China in the 1920s, wrote an essay under the title of 'Equality' in 1927:

> Not every person's natural intelligence or strength is equal. But if each person develops his mind towards service and morality...so as to contribute to the mass of humanity, then he can be regarded as equal. That is *real* equality.

Modern societies, both dictatorships and democracies, betray the ideal of equality all the time. Yet they are still committed, in their most basic rhetoric, towards a society that breaks down hierarchies and aims, even if imperfectly, towards equality as a goal. Chinese society has changed in myriad ways in the past century and more, whether it is the nature of landholding, relations between men and women, or between countryside and city, or in the duties and obligations owed by state and people to one another. This chapter examines some of these areas to ask how far Chinese society has become modern, and whether this is at odds with, or complementary to, maintaining its identity as distinctly Chinese.

Men and women

One of the most important areas where that search for equality has been fiercest is the changing roles of men and women. Mao

Zedong famously said that 'Women hold up half the sky', a rebuke to the generations of Chinese men who had regarded women as their inferiors. However, it would be wrong to take the assessment of Mao and other 20th-century revolutionaries at face value, and simply to regard the women of late imperial China as an undifferentiated, oppressed mass, or to assume that the modern era has brought Chinese women today an uncomplicated 'liberation'.

How far has the role of women changed in Chinese society since the imperial era? 'Unbound feet' has been the enduring metaphor of the change in the status of women in China between the premodern and modern eras. From the 10th century onwards, for reasons that are still unclear, the fashion developed for Chinese women to have their feet tightly bound from an early age, distorting the shape of the foot and leaving it shrunken unnaturally small for an entire lifetime. In some ways, this was surprising, as Confucian norms frowned upon the mutilation of the body. Yet the trend spread, and by the 17th century, the writer Li Yu wrote of the women of Lanzhou, in western China: 'The feet...measure at most three inches, some even smaller...Lying in bed with them, it is hard to stop fondling their golden lotus. No other pleasures of dallying with courtesans can surpass this experience.' Not all women bound their feet; it was not the custom among peasant women of the Hakka sub-ethnicity, nor among Manchu women during the Qing dynasty. But for the vast majority of women who aspired to respectable life, it was essential. A mother who did not bind her daughter's feet was doing her a disservice, for ugly, huge (that is, normal) feet would mean little prospect of a good marriage.

Opportunities for women were restricted overall in imperial China. They could not join the bureaucracy, nor were there many opportunities for them to become traders. Confucian culture did regard women as intellectually less capable than men. However, in the late imperial era, it was quite normal for well-off families to insist that women were literate. In particular, elite women during the early Qing (the 17th and 18th centuries) developed a public voice of their own in one particular area: publishing. This period coincided with a growth in mass woodblock publishing (see Chapter 6), and women writers of the era took advantage of the existence of a new readership to make their views and interests known. The Ming dynasty had seen a culture of literate courtesans become part of elite culture, whereas by the Qing, new norms of sexual control meant that chaste womanhood (widows in particular) was praised. Yet the latter change did not end the phenomenon of women writing about their own lives. The great Qing administrators Chen Hongmou and Yuan Mei were strongly in favour of literary education for women, even while affirming their belief in women's inherent lesser capabilities.

However, there was still a genuine shift in the late 19th century which saw a significant breakdown in the hierarchical relationships between women and men, in particular, poor women who would not have shared in the elite written culture of their richer contemporaries. There developed a considerable intellectual movement in favour of women's rights. Kang Youwei, one of the most prominent reformers of the late 19th century, and a Confucian in his assumptions, proposed a new society in which men and women would be equal, and marriages would operate on one-year renewable contracts. Mao Zedong, in the early days of the new Republic, published a succession of essays in Changsha analysing the fate of Miss Zhao, a young woman who had committed suicide rather than enter a forced marriage, and describing her dilemma in the most uncompromising way:

> Chinese parents all indirectly rape their sons and daughters. This is the conclusion which inevitably arises under the Chinese family system of 'parental authority' and the marriage system in which there is a 'policy of parental arrangement'.

Periodicals such as *The Ladies' Journal* and *The New Woman*, and essays on the 'woman problem' in the journal *New Youth*, were published in China's major cities. The idea of the 'new woman', autonomous, professional, and urban, was a global one during the interwar years, visible in places as far apart as the US and India. However, it was notable that in China, many of the feminist texts relating to the 'new woman' and her role were written by men. This reflected a tendency that would recur throughout the century: the role of China's women was frequently controlled by men, however sympathetic the latter may have been. In less sympathetic cases, feminism was considered a moral failing: in 1927, Nationalist activists on the look-out for leftist activists considered women with bobbed hair to be dangerously subversive and deserving of arrest or execution.

Nonetheless, the late 19th and early 20th centuries did bring about a variety of changes in Chinese society that did ultimately change the status of women. The most significant social change, and one that affected society at all levels, and in both city and countryside, was the ending of the long-standing practice of foot-binding (which was no longer practised by the 1920s). Beyond the ending of foot-binding, however, the majority of social changes for women in pre-1949 China were concentrated in the city. As the concept of the 'New Woman' became well known among urban elites, the character of Nora, from Ibsen's play *A Doll's House*, became a powerful role model: Nora leaves home at the end of the play, abandoning her husband and children to find an independent role for herself. Yet society as a whole changed less quickly than the aspirant Noras would have wished. The writer Lu Xun asked the pointed question 'What happens after Nora leaves home?' Even so, there were opportunities that had simply not been there a few decades before. Women became journalists, lawyers, and students: education, particularly at university level, was the preserve of a very small number of people of either sex, but university radicals were at the heart of feminist thought of the era.

Even mainstream politics found a role for women. The Communist movement made great play of its commitments to gender equality from its earliest days. The reality was not always so clear: Xiang Jingyu, the most prominent woman in the CCP in the 1920s, found that her feminist concerns were repeatedly sidelined so that the party could accommodate the prejudices of rural men, who were its primary target at that time. Although the Nationalist Party, in government from 1928, did little fundamentally to challenge gender roles, it did give women citizenship rights and (theoretically) equal rights to status in marriage and inheritance. Poorer urban women found opportunities in the massive social change of the era. As capitalist modernity brought factories to China's cities, in particular Shanghai, rural women were recruited from the countryside to work in the factories that made the silk and cotton cloth that underpinned the textile industry. These jobs were backbreaking and dangerous, and paid little: but they marked a move for ordinary women from the confines of the village cottage industry, linked to their families, to a more autonomous, urban way of life. For women still in the countryside, however, change would be much longer in coming.

The single most devastating event of China's 20th century, the Sino-Japanese War (or War of Resistance to Japan), changed society from top to bottom. This included the destabilization of the settled village society that still dominated rural China. In the areas controlled by the Communists, the conditions of wartime were used to create radical social movements that would eventually reshape China in the post-war era, and in which traditional gender roles would be further undermined.

In 1949, the CCP's victory 'overturned heaven and earth'. The following decades would be an ambivalent time for Chinese women. The Great Leap Forward ended up causing China's most devastating famine of the 20th century, and it is this disaster by which its effects must ultimately be judged. However, it also marks the period when women's roles were most notably equal. The Cultural Revolution, in contrast, offered a much more ambivalent view of womanhood. Although it stressed social equality and took delight in breaking down boundaries, the Cultural Revolution's stress on violence and radical change swiftly made it clear that it took masculine values as its default. Women were shown in the vivid Socialist Realist posters of the era wielding rifles; men were not shown feeding babies. In addition, the unbearably close attention that was paid to people's private lives during the Cultural Revolution led to prurient accusations, often against women, every bit as censorious as those of the premodern era which Mao had condemned. Even so, there were more women party officials and individually paid workers in the economy than at any other point in Chinese history.

The era of reform after 1978 saw a loosening of society in all areas. For women and men alike, some of the claustrophobic restrictions of the Cultural Revolution were removed. Romance was no longer a sign of bourgeois backsliding; nor was wearing stylish clothing or make-up. Yet the social egalitarianism of the earlier era gave way in part to a type of paternalism familiar from the past. Within the party, the number of women party officials, both at local and elite level, dropped dramatically. In the labour market, women often found it harder to gain jobs because their prospective employers assumed they would swiftly leave to have children (and did not wish to make provision for this eventuality). The CCP did a significant amount to enforce the rights of widows and daughters to an equal share of inheritance, but even today, especially in rural China, premodern assumptions that males have priority in inheritance remain powerful (even though women could inherit property in some circumstances in late imperial China).

Most notable of all, the era saw the arrival of the 'one-child policy'. Worried about the exploding population growth rate, in 1979 the government placed severe restrictions on the number of children that a family could have. Although the regulations changed somewhat from their initial version over the years, the general rule was that one child was permitted for an urban family, and that rural families might have a second child if the first was a girl. The policy was justified on demographic grounds. Yet it also put a severe question mark over the government's commitment to equality. First, the onus was put squarely (if not explicitly) on women to prevent pregnancy, making it a personal, biological responsibility from which wider society, and men in general, were excused. Then, permission for rural families to try again if the first child was a girl effectively admitted that the old, Confucian hierarchies in the countryside were unchangeable, and that the desire for a boy first and foremost should be indulged. The mixture of old attitudes and new technology in China led to unforeseen and potentially damaging effects. The use of ultrasound to find out a child's sex before birth has led to large numbers of abortions of female foetuses, leading to a severe gender imbalance in parts of China. The policy had another effect that will only be seen in future decades: China is rapidly ageing. Around 140 million Chinese are elderly (that is, about 10 per cent of the population) in the early 21st century; by 2025, UN estimates suggest that there will be 326 million Chinese over 50 and fewer than 278 million under 20. For these reasons, in 2015, the policy changed to allow two children for any couple who wished to have them.

War and society

Confucian norms stressed the importance of harmony and order. This was an understandable reaction to the times that Confucius lived in, the era of the Warring States when kingdoms wrestled with each other for supremacy. The terrain of China has been marked, since the 19th century, by war and conflict, the literal battles often shaping the metaphorical ones which have influenced the wider society. The experience of almost constant warfare shaped China from the Taiping Wars of 1850–64 until the establishment of the People's Republic in 1949. Even then, Chinese society was not truly at peace for decades.

The Qing dynasty, from its foundation in 1644 to its zenith in the 18th century, took part in massive expansion of Chinese territory to the west and south. These were wars of conquest, however, rather than being internally disruptive civil wars or foreign invasions. Just as Britain fought abroad for colonies in the 19th century while its metropolitan societies remained peaceful and prosperous, the High Qing era did not see conquest linked with domestic collapse.

The wars of the mid-19th century onwards were very different. Late Qing China had a variety of problems that would have come to prominence anyway, such as a major agricultural crisis and an outstripping of the state's ability to collect revenue. However, the impact of war hastened the state's collapse. The famous first Opium War (1839–42) did not in fact cause much social instability in its own right, as it was fought mostly on the coast and at sea: its importance was in demonstrating the reversal of power between the Qing empire and the British. A series of later wars caused much greater direct impact, however, such as the Taiping War of 1850–64. Warfare did a great deal to corrode Qing authority. In physical terms, it destroyed the agricultural base in some of China's richest areas, further reducing the state's ability to generate revenue. In psychological terms, it enforced the idea that central state authority was no longer strong enough to protect people from attack. Local elites, many of whom had been instrumental in forming local militias which had eventually helped defeat the Taiping, became a new source of power that became an alternative to the government in Beijing. Eventually, that local power would undermine the Qing fatally in 1911.

In fact, the dynasty's downfall in 1911 was less violent than many revolutions, although there were notable atrocities such as the murder of Manchus by Han Chinese revolutionaries in cities such as Nanjing. However, the new Republic of China was wracked by war throughout its existence, so much so that many reformers summarized the country's troubles as 'imperialism from outside, warlordism from within'. Civil wars wracked the country from 1916 onwards. The establishment of the Nationalist government in 1928 was supposed to bring the period of warfare to an end. But in practice, Chiang Kaishek's government was still at war through most of the following years: with the Communists, with rival militarists, and then with the Japanese. These wars were not just passing phenomena: peasants found their crops confiscated or destroyed, and faith in central authority remained scanty.

It did seem possible by the mid-1930s that Chiang's government was on the way to consolidating its power. The Communists were on the run after the Long March, and Chiang had managed to establish an uneasy truce with most of the regional militarists who had acted against him. But the outbreak of war with Japan in 1937 put paid to any hope of modernization under a centralized, stable state. Reform required peace, stability, a reliable tax revenue stream, and access to international and domestic markets. The Nationalist government had none of these. The result was a state that turned in on itself. Corruption, black marketeering, and runaway inflation in the Nationalist zone led to a collapse of trust in the government, paving the way for the Communist victory in the Civil War which followed in 1946–9. The poor record of the Nationalist government in the last years of the war has rightly been blamed for the loss of support from the Chinese public. However, this explanation cannot stand alone without an understanding of quite how thoroughly the war against Japan had destroyed the basis of the society which the Nationalists governed. On the other hand, one undeniable change did happen in both Nationalist and Communist areas as a result of the war: the state became entwined with society much more inextricably. The political parties demanded that their people support them in their struggle against Japan; in turn, refugees demanded food, and ordinary citizens expected protection from air-raids, and more widely, the establishment of a society that would reward them for the sacrifices they had made during the long years of war.

Mao's victory in 1949 has generally been regarded as the end of the period of war in China, as the country was finally united under a single government. Yet the organization of Maoist China was, in significant ways, 'war by other means'. The Cultural Revolution saw pitched battles in the streets of cities such as Shanghai and Chengdu, and the feud with the USSR saw society and economy placed on a war footing in case of invasion. Not until the reform era did society return to something genuinely like peacetime as understood in most liberal societies. It is understandable that so many Chinese live in fear, above all else, of *luan* (chaos).

The war against Japan has come back to haunt China in recent years. Since the 1980s, the search for a new nationalism that can help to boost the CCP's legitimacy as a party representing all Chinese, as well as promote reunification with Taiwan, has led to a new emphasis on the history of the war years. During Mao's period in power, the war against Japan was downplayed in history books: the Nationalist contribution to victory could not be mentioned, China had little interest in provoking a now pacified Japan, and therefore little public attention was paid to Japanese war crimes in China. From the 1980s, however, the post-Mao regime has stressed that the war against Japan was a moment of great suffering but also of renewal for China. A museum commemorating the Nanjing Massacre was opened in that city in 1985 (nearly half a century after the event itself), and films and books began to give a more balanced and thoughtful account of the contribution made by the Nationalists to the defence of China. The period around the 70th anniversary of the ending of the war against Japan, in 2015, saw numerous reminders to the Chinese public of the significance of the wartime experience, including books, television programmes and video games, along with a major parade in Tian'anmen Square on the new public holiday of 3 September (timed for the day after the anniversary of the official Japanese surrender on 2 September 1945).

Is China a richer society than under Mao?

Significant progress in poverty reduction has taken place in China during the reform era. In 2001, infant mortality rates in China were 31 per 1,000 live births (as opposed to 7 in the US and 67 in India); life expectancy was 70.6 years (compared to 76.9 in the US and 63.3 in India). In the same year, 93 per cent of primary-school-age children were attending school (95 per cent in the US, 83 per cent in India).

There are complexities beyond the headline figures, of course. A survey of data in 2012 revealed that the bottom 25 per cent of China's population controlled just 1 per cent of its wealth. Nonetheless, the Chinese achievement is impressive and has done a great deal to create personal, if not political, freedoms that come from greater economic opportunity and well-being. Much of the countryside has, of course, become richer between the mid-1980s and mid-2000s: during that time, urban incomes increased by 14.1 per cent whereas rural incomes increased 11 per cent. But the fastest-growing group in society is the urban middle class: in 2012, some 68 per cent of all urban households were judged to be middle class, defined as having a household income of between Y60,000 and Y229,000. Furthermore, popular aspirations are phrased in terms that make it clear that the old, agricultural China is regarded as the past, not the future: a college education, a pleasant city apartment with running water and flush lavatory, consumer goods (electronic equipment, white goods, a car), and services (leisure travel, cable television). New consumers are emerging elsewhere, for instance among the large population of migrant labourers within China (more than 250 million in 2012) from areas of poverty to richer regions, where they seek work in under-regulated industries such as construction.

China is undeniably a richer society in per capita terms than it was under Mao (see also Chapter 5). However, it is also much more unequal. Services such as free health care, education, and guaranteed employment that were part of the Maoist social contract were abandoned during the reform era, and increased cash incomes have still often not been enough to compensate for the new, free-market costs of hospital treatment, medicines, or local school fees.

Is China free?

China is often portrayed, in the present day, as the most prominent example of a society that is not free. Since 2014, there has been a significant clampdown in China on websites such as Google. Yet there clearly *are* other freedoms that have made a real difference to the lives of Chinese at all levels of society.

To understand the significance of these freedoms, one must ask what concept of freedom had previously existed in premodern China. In late imperial China, the state was widespread but relatively shallow. Local magistrates, provincial governors, and bureaucrats of various sorts kept the network of the empire running, but its active reach into the lives of ordinary Chinese was much less strong than the increasingly intrusive state of the 20th century, which reached its apogee in Mao's China. For poor rural farmers, just as for the poor of London or Paris, there was little freedom to act, because economic deprivation limited the scope for action. Yet as in early modern Europe, there was also less active state interference in people's lives than in the 20th century.

Economic freedom helped shape premodern China, with a sophisticated market economy allowing land to be bought and sold, and the ability to accumulate capital allowing business ventures to form and prosper, even though the economy was not mechanized or industrialized in the way that its European equivalent was. People and goods were generally free to travel around the empire, although the trade in certain key goods such as grain and salt was heavily regulated.

However, the freedoms of political action which are associated with the aftermath of the English, American, and French revolutions are not so easy to detect in premodern China, just as they would have been hard to find in much of Central and Eastern Europe, Russia, and the Iberian peninsula at the same time. The education and immersion into elite values that characterized scholars and officials also made them subject to the moral rules of Confucian governance, and in particular, their duty to speak out when injustice was done by those in power. Nonetheless, the Chinese system did not institutionalize protection for those who spoke out in this way and that could make open dissent a morally virtuous but personally perilous undertaking.

The usage of law in modern China has been one way to assess how free the society has been. The Republican state was much more tied to the European idea of constitutions and codified law than the Qing empire had been, because its structure was so heavily influenced by Western and Japanese models. The Republic was not a strong state, even under the Nationalists, although it was more stable and promising than later historiography has tended to acknowledge (see Chapters 2 and 3). Ironically, though, this failing was a blessing for personal freedom. Unlike the Qing, the modern Republican state desired strong day-to-day control over its people; but it had insufficient resources to exercise that control.

Nonetheless, freedoms were restricted in the Republic, and became more so under Mao. Even in the era of reform, the 'cage' has been opened only part of the way. China still has very large numbers (probably thousands) of prisoners held essentially for political offences, such as attempting to set up a new party, joining a banned religious group, or placing dissenting views on a weblog. There are highly credible reports that such prisoners are treated roughly and even tortured. However, China today is overall not a totalitarian state, nor a military junta, nor a state run at the personal whim of a dictator. From a Western point of view, China even appears (perhaps deceptively) free compared to, say, many societies in the Middle East. To use Isaiah Berlin's terms, 'positive liberty' in China today is highly restricted—there is no freedom to establish rival political organizations, the media is highly circumscribed and censored, and public protests, although common, are usually shut down fast. But 'negative liberty'—the freedom, essentially, to be left alone by the state in matters of personal choice—is undeniably strong. The Chinese of today are free to set up businesses, wear clothes as they please, buy consumer goods, and travel (though not live) where they wish within China and even go to many places outside it. These freedoms are restricted by economic capability and by corruption, which prevents a truly free set of choices being made. Nonetheless, these freedoms have a real impact. In the Cultural Revolution, wearing leather shoes or Western hairstyles could lead to attacks from the Red Guards. Markets were almost entirely controlled by the state, making events such as the Great Leap Forward possible. The freedoms such as the growth in market information that prevent such events happening again, are less romantic than the inspiring 'Goddess of Democracy' statue hoisted by students in Tian'anmen Square in 1989. But they are important and their enactment has not been without risk for the CCP. This is not to accept at face value the CCP's self-serving argument that only it can decide how far and how fast China can safely liberalize. But today's China relates to its people in a very different way from the state under Mao.

One of the major factors that has marked the growth of freedom in post-1978 China is exposure to the outside world. Before 1949, China was highly internationalized, and its modernity was shaped by constant interaction with the outside world (often in the unwelcome form of imperialism). The Mao era saw China look inward more and more, and even the USSR was unwelcome after the Sino-Soviet split of the 1960s. Since 1978, however, China has embraced the outside world with enthusiasm. The Tian'anmen Square tragedy of 1989 looked as if it might put a stop to this process, but in fact it proved only a temporary obstacle. In the early 21st century, the Chinese are once again a globalized people in their homeland as well as around the world. Chinese students are among the largest communities in the universities of the US, the UK, and Australia. Chinese tourists are a common sight in Bangkok, Paris, and London. Chinese academics regularly visit the West for conferences, and Chinese businessmen do deals on six continents, in recent years finding new opportunities in Latin America and Africa. In the 1980s, there was an expectation that many of the Chinese students who went to the West to study would choose to stay there. Now it is far more common to find entrepreneurs whose ambition is to return and set up a firm in China's burgeoning market. Books by Chinese business gurus are piled up in the major bookshops of Beijing and Shanghai. For the majority who cannot yet afford to travel abroad, there are endless television news reports and documentaries on societies in other parts of the world. Foreign television shows are imported and dubbed; likewise foreign films. China is not isolated from the outside world. The arrival of the internet, for instance, has become a very important part of the ability of China's new middle class to engage with the outside world. Many sites, including the BBC Chinese service and sites relating to the Tian'anmen Square killings in 1989, are blocked from within China. But thousands of others, relating to foreign films, university courses, news stories, celebrity gossip, and corporations are not. Many Chinese understand perfectly well the freedoms available in other parts of the world, but they choose not to embrace them—or at any rate, not to embrace them yet.

The wider Chinese world opens up intriguing divisions between what is 'free' and 'democratic'. China itself is neither fully free nor democratic. Taiwan, since the 1990s, has been both free and democratic. Singapore, a largely Chinese society, is democratic, in that it has regular elections which are nominally open to opposition candidates (but at high cost to themselves), but is not free (the media and political activism are both heavily regulated).

Most intriguing, though, is Hong Kong, which is little more democratic than it was under the British. Yet it is a very free society: although there is growing and worrying political pressure and self-censorship, it has a lively press, it is still common to see books attacking the Chinese government, and it supports a variety of political parties (although the legislature is arranged to prevent any overly liberal party ever coming to power). There are few, if any, other such free, undemocratic societies.

One particular freedom that has been vigorously exercised since 1978 is the new freedom to practise religion. Officially, religious freedom was always guaranteed under the People's Republic of China's constitution, but during periods such as the Cultural Revolution, religious practice was condemned as superstition, and was politically dangerous. However, in the present day, the state feels that mainstream religion acts as a social glue, and no longer seeks to discourage it, instead recognizing state-approved versions of Taoism, Buddhism, Islam, Protestantism, and Catholicism. Christianity has become more widespread (one survey suggested there were some 100 million Christians in 2014) and has found a particular constituency among young urban Chinese, who associate the faith with modernity. Around 1.5 per cent of the population of China is Muslim. However, the state is deeply distrustful of religious movements that seem to offer any organization that challenges the government, or even seek merely to avoid its surveillance: the Falun Gong movement is one of the best known of such groupings.

The passion for improvement

What ties together the themes that characterize Chinese society? Since the violent, transformative impact of modernity on Chinese society from the 19th century onwards, it is clear that China has not been a 'Confucian' society in the premodern sense. However, significant cultural influences have remained from that earlier era. Perhaps one of the most powerful, and one whose influence can be found in a wide variety of areas, is the idea of *xiushen*, or personal cultivation: the Confucian idea that the way to living a decent and ethical life lay in improving the self, with the aim of becoming a *junzi* (gentleman, person of integrity) or *sheng* (sage). Education is a clear and obvious means of achieving that goal, but contemplation and reflection also make up a large part of it.

Although the modern Chinese states of the 20th century tried to distance themselves from what they saw as the backward elements of the Confucian tradition, the idea of *xiushen* as a collective, as well as a personal, goal persisted strongly. In the 1930s, as part of a drive for national renewal and in a bid to undermine Communist ideology, Chiang Kaishek launched the New Life Movement, which attempted to drill the Chinese population into regulating its behaviour (see Chapter 3). However, despite its failure in the 1930s, there are strong and ever-more-explicit echoes of the New Life Movement in contemporary China, where points are given by local committees to residents who throw away their garbage and put out plants to decorate their houses. In the run-up to the Olympics in 2008, Beijing residents were told of a new 'morality evaluation index' which would give credit for 'displays of patriotism, large book collections, and balconies full of potted plants' and lower grades for 'alcohol abuse, noise complaints, pollution, or a violation of licences covering internet cafes and karaoke parlours'. Public toilets in tourist areas were also upgraded and star-rated (see Figure 9 and Box 3).

The idea of *suzhi* has also taken hold in recent years, a word hard to translate but usually rendered as 'population quality'. To some, it has a faintly eugenic air: educated Chinese will often claim that peasant *suzhi* is lower than that of people in the cities. Yet an individual's or group's *suzhi* is not permanently fixed, and education is one means to improve it. The legacy of *xiushen* can be seen even through the pseudo-scientific language that surrounds *suzhi*. This debate also reflects a wider disparity developing in China. During Mao's period of rule, his policies (in particular, the Great Leap Forward and Cultural Revolution) were designed to

9. Beijing geared up for Olympic fever after being awarded the Games in 2001. Here a worker carries an Olympic Rings decoration made of waste tyres from a car-wash.

Box 3 Olympics 2008

In 2001, two Chinese obsessions came together: sport and respect. The International Olympics Committee (IOC) announced that the 29th Olympiad of the modern era would be held in Beijing in 2008.

The 20th century was marked by the use of sport as an indicator of national prowess. Premodern Chinese culture, however, had not regarded physical exercise as the mark of manhood, but rather, praised the ideal of the Confucian gentleman, learned and separated from the world of exertion. This changed in the late Qing and Republican periods. Many thinkers, influenced by Social Darwinist ideas, felt that China's devotion to scholarship over physical prowess was leading the nation to destruction. Mao Zedong, in his earliest writings, noted that 'Physical exercise should be rude and rough', and even sketched out a personal exercise plan involving buttock thrusts.

Nor was Mao alone. China's leaders were aware that its weak international status meant that it had a long way to go to exercise military power in its own right. However, there were other forms of cultural power that could be shown off in the international arena. China sent a team to the Olympics for the first time in 1936, that year held in Berlin, perhaps the most purely political Games that the Olympic Movement has ever seen. The Cold War was also a sporting war, with US–Soviet rivalries played out every four years. China did not compete until the 1980s, but then quickly made its mark in sports such as gymnastics. By participating in the Olympics, China was making a clear indication that it was once again part of the world community.

The same motivation fuelled the decision in 1992–3 by the Chinese government to bid to bring the millennium Olympiad to Beijing. Huge amounts of money were spent in the bid, and the result was piped through loudspeakers live into Tian'anmen Square. However, the shocked crowd heard that the lucky city was to be ... Sydney. The unstated, but widely felt, subtext was that the IOC could not award the Games to China just four years after the killings in Tian'anmen Square. The international world of sport could not give China permission to slip back into the community of nations as if nothing had happened. However, the 1993 snub proved to be a start, rather than an end, to the process. Within two years, China held a major international event, the 1995 UN Women's Conference. Finally, in 2000–1, Beijing once again put itself forward as a candidate city. This time, it started as the favourite and never lost the position. By the time the Games started, the killings in Tian'anmen Square would be almost twenty years in the past. The China that had been given the Games was no longer a shell-shocked society recovering from an internal social crisis, but a confident regional power with a global reach.

(continued)

Box 3 Continued

For many countries, such as Italy, Germany, South Korea, and Spain, the holding of the Olympic Games was a symbolic 'coming out' from a dictatorial past. The message of Beijing 2008 was more ambiguous. Certainly, the Games symbolized China opening itself up to the outside world, a welcoming host to the family of nations. Yet Beijing 2008 was the first Olympics held since Moscow 1980 in a country that is not democratic. And in the years since then, several other non-democratic societies have been awarded prestigious events such as the Winter Olympics and the football World Cup. The Beijing Olympics certainly symbolized China's arrival in the international community; what is more, they did so largely on China's terms, and in ways that validated authoritarianism well beyond China. In 2015, the Winter Olympics were awarded to Beijing, the first time a city has hosted both Summer and Winter Games; the only competitor to host them was Almaty, in equally illiberal Kazakhstan.

attack, with great violence, the traditional air of superiority that urban-dwellers, and educated Chinese in particular, displayed towards the rural population. (Mao's introduction of a household registration system to prevent farmers from migrating to the cities undercut this intention, of course.) However, the trend has been firmly in the other direction in recent years.

The other Chinas

This book has mostly been concerned with events on the mainland. However, the main landmass that we know as China has always been affected, and continues to be so today, by Chinese societies well beyond its own borders.

Taiwan remains the major piece of 'unfinished business' from the Cold War, in the eyes of the Beijing leadership. After the return of

Hong Kong to Chinese rule in 1997, reunification with Taiwan moved much further up the political agenda in the rhetoric of the mainland. At the same time, movements within Taiwan itself have given fuel to the idea that the island should declare independence. It remains a commonly held point of view among mainland Chinese that the country will only have full territorial integrity when Taiwan is part of the 'motherland' once more, and that all means, including war, are acceptable to bring about this aim. China's claim to Taiwan dates from imperial times, when the island came under the control of the Qing dynasty. Yet this interpretation of history hides complications. The Qing empire also contained what is now Outer Mongolia, yet there are no calls for that country to be reabsorbed into China. More crucially, Chinese rhetoric does not call attention to the fact that, since the late 19th century, Taiwan has only been part of a united Chinese state for four years (from 1945 to 1949).

Taiwan's history is a complex one. Its earliest inhabitants were Malayo-Polynesian aboriginals, and ethnic Chinese settlers arrived there in significant numbers only from the 16th century. The Qing dynasty did incorporate Taiwan into its territory, but as late as the 19th century, the island was regarded even by the Chinese as a remote, frontier place.

The island did not have long to develop under the Qing. The dynasty, having lost the Sino-Japanese War of 1894–5, was forced to cede the island to Japan, making it the first formal Japanese colony in Asia. For the next half-century, Taiwanese grew up under Japanese colonial governance. While that rule was patronizing and often harsh, particularly during the war years of the 1930s and 1940s, it was not marked by the brutality seen in Japanese behaviour on the mainland. The Japanese succeeded in creating an indigenous colonial elite, who spoke Japanese as easily (or more so) than Mandarin or Taiwanese Chinese, and who even today, regard their former colonial rulers with ambivalence or even tolerant approval. Former president Lee Teng-hui once declared that he 'was a Japanese before the age of 22' (that is, until 1945). What is also notable is that the turbulent history of China's Republican period—the 1911 revolution, May Fourth Movement, the Nationalist government, and above all, the rise of a powerful Communist Party—had little relevance for the island. Even the most traumatic event of all, the war against Japan, happened off-island, and some Taiwanese fought on the Japanese side, outraging their mainland compatriots.

Japan's defeat in 1945 also saw the island handed back to Chinese (Nationalist) control, and the islanders' view of their colonial experience became rosier in retrospect because of the harsh nature of what followed. The Nationalists treated Taiwan as if they were an occupying power, rather than liberators. As the Civil War raged on the mainland, the government clamped down yet further on dissent in Taiwan. On 28 February 1947, an altercation between the police and an old woman selling contraband cigarettes spiralled into mass protests by the island's indigenous population against Nationalist rule. The reaction was swift and brutal, with thousands of Taiwanese being killed or imprisoned. Any discussion of the event was banned, particularly after Chiang Kaishek fled to the island in 1949, and '2-28', as the events became known, lay under the surface of the island's memories for decades.

After 1949, the island lived in a state increasingly detached from reality as the 'Republic of China' in exile, waiting to take the mainland back from the Communist rebels and bandits. Military rule was declared, and the Nationalist rulers continued to commit human rights abuses well into the 1980s. Yet although political dissent was strongly repressed in Taiwan, a very successful economic model was being laid out at the same time. The Nationalists tackled one of their great failures on the mainland, the reform of land ownership, a project heavily encouraged by American advisers. They also used governmental power to create an economy driven by exports, in particular consumer goods, leather, wood, and paper. During the 1960s, Taiwan's exports rose by around 20 per cent each year, and the economy as a whole grew by 9.2 per cent annually.

Most notably, Taiwan moved firmly towards democracy. The 1970s saw new interest groups emerge, including the *Tang-wai* ('outside the Party') movement on behalf of which Chen Shui-bian had stood for office. Many of these interest groups advocated more power for the local population of the island, whose interests had been subsumed by the agenda of the 'mainlanders' who had fled to Taiwan in 1949. Chiang died in 1975 and was succeeded by his son, Chiang Ching-kuo, who eventually took the decision to legalize dissent. By the 1990s, Taiwan had a genuine liberal democracy: in 2000, for the first time, the presidency was won by a representative of the opposition, the former dissident, Chen Shui-bian. Eight years later, beset by scandal, Chen lost office in turn to his Nationalist challenger, Ma Ying-jeou.

Taiwan's system diverged from that of China at the same time that the relationship across the Straits was strengthened in a variety of ways. As China opened up under Deng, 'Taiwan compatriots' were encouraged to visit the mainland and invest in it. At the same time, the CCP became ever more concerned by the number of Taiwanese who no longer considered reunification with China desirable or even relevant, and the mainland made threatening noises about its right to invade Taiwan in the event of a declaration of independence.

Yet China has been highly successful in convincing the world of its own position on Taiwan: that Taiwan independence is unthinkable. This is in some ways odd, because Taiwan has not always been such a polarizing issue: it was not central to the territorial rhetoric of the Republican governments, or during much of Mao's period in power. Nor is reunification necessary for China's continued economic growth or political influence. Nor, any longer, does the mainland demand that Taiwan become like China: the terms offered for reunification since 1978 have become something like a reunification in name only, involving Taiwan maintaining its own political system and even military.

In the 1980s, it was Hong Kong that was more prominent in the news. In 1898, the British had forced China to grant a 99-year lease on the New Territories that bordered Hong Kong Island and Kowloon, which had been seized after the Opium Wars. As the lease began to end, Deng Xiaoping's government made it clear that they were determined to take Hong Kong back, but also gave assurances that the colony's way of life would not be altered for at least 50 years. In 1984, a Chinese–British accord formalized the agreement for handover, but two events made the final days before the handover in 1997 much less calm than anticipated: the Tian'anmen Square demonstrations of 1989 and the appointment of a democratic politician, Chris Patten, rather than a civil servant, as the last governor of the territory. Patten introduced much wider voting rights in Hong Kong's highly limited elections, infuriating Beijing, who regarded the act as a breach of the spirit if not the letter of the handover agreement. There was much speculation within Hong Kong as well as from business and political circles in Britain that the reforms would harm Sino-British relations for years to come.

In fact, although the wider franchise was abolished as soon as the Chinese took over, neither Chinese rule nor Patten's reforms immediately had the dire effects that some feared. Hong Kong continues to have a strongly international feel, a free business environment, and a broadly free media. Most notably, there is a strong popular interest in politics, partly because there are growing fears within Hong Kong about the direction of political travel. A widespread student movement occupied parts of the city in the autumn of 2014, protesting primarily about the attempt to impose a very limited system of election for the territory's Chief Executive from 2017 (allowing Beijing to select candidates who would only then be subject to a popular vote), but also drawing on young people's anger about increasingly limited economic opportunities in the city.

The 'overseas Chinese', that is, the ethnic Chinese diaspora living outside the boundaries of China itself, have always been an important part of the story of China's development. There are distinct communities: in the 19th and early 20th centuries, coastal Chinese moved to California, to South Africa, and to Britain. In the post-1945 period, there was a new influx of Hong Kongers to Britain, and after 1965, when racial quotas were lifted, of Hong Kong, Taiwan, and South East Asian Chinese to the United States. From the late 1970s, it once again became possible to emigrate from mainland China, and in the decades that have followed, it has become straightforward for those who wish to emigrate, and can afford it, to do so. Migration has now become truly global and is a commercialized entity with agencies including local governments and (sometimes criminal) 'snakehead' gangmasters all playing a role in the diffusion of Chinese to North America, Europe, Australasia, Africa, and Latin America. Migrants also seek education, no longer being content to remain in relatively low-skilled industries (such as restauranteuring) that distinguished them in the 1960s and 1970s.

A modern society?

Overall, there are plenty of aspects of contemporary Chinese society that are directly connected to the world of a hundred or five hundred years ago: popular religious practice, the preference for male children in the countryside, and the stress on hierarchy among them. Yet other aspects of society could only have been formed in the modern world: the presence in a globalized economy where labour as well as capital has become more mobile, the stress on the language of equality and rights that comes from a century of nationalist and communist politics, and the notion that the state and the people are, and should be, closely entwined with one another, the latter tendency strengthened by the experience of total warfare, whether against Japan or against 'class enemies'. Modern Chinese society is both Chinese *and* modern.

Chapter 5
Is China's economy modern?

In December 2014, the International Monetary Fund (IMF) reported that the Chinese economy was worth $17.6 trillion, slightly exceeding the $17.4 trillion estimate for the US. This marked the first time since 1872 that the US had not topped the list of the world's largest economies. One should note that these figures were adjusted for Purchasing Power Parity (PPP), without which the size of China's economy would seem much smaller, and most measures do still categorize China as the second biggest rather than biggest global economy. Yet the IMF statistic highlighted the paradox of China's status in the 21st century: economic superpower that moves the world's markets, impoverished developing country—or both? China's economy grew at around 10 per cent per year in the first decade of the 21st century, one of the highest sustained rates of wealth creation in world history. The rate of growth seems all the more astounding because it is contrasted with the relatively inward-looking Mao period and the era of war that preceded it. Yet it was not entirely unprecedented: the growth in the Chinese economy marked something of a return to the early modern era, when the Chinese economy was comparable with those of Europe. The 20th century was undoubtedly a troubled time for the Chinese economy, particularly for its huge agricultural sector, which was devastated by war, depression, and the impact, within a short number of years, of not enough government (under the Nationalists) and too much (under the Communists).

The characteristics of a modern economy include an active enthusiasm for growth, capital investment and industrialization, and ever-increasing productivity through the development of technology. In those terms, China's economy, particularly since 1978, measures up spectacularly. Among the key factors have been Foreign Direct Investment (FDI), cheap labour, and the immense importance of overseas Chinese financial and human investment, as well as continuing investment in education and scientific and technical research and development. However, there have been real drawbacks to this growth. In particular, the rise in Chinese consumption and the impact of unrestricted production on China's environment have stored up immense, and expensive, problems for the next generation of Chinese leaders.

The origins of the modern Chinese economy

To trace the origins of China's engagement with the modern, globalized economy, we need to go back over a thousand years. The Song dynasty (960–1276) saw one of the major changes in the Chinese economy. Up to that point, the majority of farmers were subsistence cultivators, but by the end of the dynasty, they had become specialists, producing cash crops or making goods for the market. At the same time, a countrywide internal market developed in China, which would go on to expand under the Ming and Qing. A paper and metal money economy also thrived during this period, leading to financial crises and increasing contrasts between the very rich and very poor. Yet overall, it was clear that this was a time of immense growth for the Chinese economy, and it was certainly comparable with the increasingly commercial nature of the European economy of the time.

However, half a millennium later, events in the economies of Europe and China would differ greatly. China's economy essentially produced more of the same. The 18th century in particular was a golden time for China. Its territory expanded as the Qing conquered lands to the west and the north. The introduction of

New World grains in previous decades had expanded the number of crops that could be grown on land that had previously been considered barren, allowing the population to spread and grow. By the end of the 18th century, China's population had doubled from 150 to 300 million people.

The same period, of course, saw a revolution in Europe, starting in England: the agrarian and industrial revolutions of the modern era. The historian Kenneth Pomeranz has been particularly associated with a specific question: Why was there a 'great divergence' between Europe and China in the 18th century? He argued that Europe's most advanced area (England) and China's (the Yangtze valley) were at a comparable level of development in around 1800. Why, then, was it England that saw unprecedented, dynamic growth? His argument was based on a number of factors, but principal among them was that England had benefited from conveniently sited coal mines and colonies, neither of which was available in the Yangtze delta at the same time. While there has been lively debate about specific details, there is now considerable consensus that it is meaningful to compare the economies of early modern England and China, in terms of their relative stage of economic dynamism and growth.

Nonetheless, the 'great divergence' did take place. China developed an ever-more commercial economy in the late imperial era, but until the late 19th century, it did not develop the type of industrial modernity seen in the West. This changed, inevitably, with the advent of Western imperialism in the mid-19th century, which brought about profound changes both to China's industrial and agricultural economy.

The collision with Imperialism and industrialization

The problems of rural China were not just of Western making. By the late 18th century, signs of an agricultural crisis in the Chinese countryside were evident. Over a century later, in the 1930s, the

British economist R. H. Tawney was moved by the plight of China's peasants to declare: 'There are districts in which the position of the rural population is that of a man standing up to the neck in water, so that even a ripple is sufficient to drown him.' For years, it seemed self-evident that the rural economy in China was a disaster in the period before 1949. Yet over the past 20 years or so, serious reassessments have been made. These have led to strongly opposing views as to causation, but the old view that China's agricultural crisis was bad and getting worse all the way into the mid-1930s is highly misleading.

One of the major reasons that China's peasant economy in the century from the 1840s to the 1940s has come in for such criticism is the turbulent nature of the times. However, economic historians no longer make the assumption (common enough from the mid-20th century until around the 1980s) that tumultuous politics necessarily led to disastrous consequences for the Chinese economy. In fact, there is significant evidence that, after the late Qing crisis, the overall agricultural economy of China became more productive and profitable. The historian Loren Brandt argued that 'between the 1890s and 1930s, agricultural output in Central and East China increased more than two times the estimated population growth of 0.6 percent per annum'. He attributed this to various factors. One was increasing specialization, including the growing of cash crops such as cotton. Commercial and technological factors also changed things: by the 1930s, more than 40 per cent of all farm households in that region were using commercial fertilizers, and rural credit was becoming easier to obtain. Seed development, including higher-yielding varieties of rice, were also coming into usage. Yet the economy remained a very poor one. In the mid-1930s, per capita GNP was only 60 yuan (around US$200).

China had been significantly involved in trade within Asia for centuries, and had played a major part in the luxury goods market in Europe in the 18th century (providing tea, porcelain, and silks in return for silver). But China only entered the international market fully in the modern era. This was a dubious blessing, because that international market was in large part a function of imperialist attempts to open China up. Nonetheless, China's semi-colonial status did give it a level of economic autonomy that a true colony, such as India, did not have, particularly after the Nationalists managed to regain autonomy to set tariffs (taxes on imports) by the early 1930s.

The Sino-Japanese War would cripple China's economy, just as it destroyed China's state-building experiment of the same era. Not enough research has yet been done on agricultural production in wartime China, but there can be no doubt that the war dealt a heavy blow to the progress made in the decades before 1937. China's transport networks were destroyed, and large parts of its agricultural land were laid waste by war. Not everything was lost: good harvests in the first year of the war meant that the breadbasket province of Sichuan was able to supply the areas of China under Nationalist control, and the government continued experimentation with fertilizers and new seed varieties. But wartime did not provide anything like normal conditions to assess economic progress. The final years of the war saw the economy collapse. Scarcity of consumer goods led to black marketeering and hyperinflation. The loss of important agricultural areas to the Japanese during the offensives of 1943–4 led to widespread starvation in the countryside and increased the population's alienation from the Nationalist government. Even after the war was over, the financial crisis continued throughout the 1946–9 civil war, and Chiang's government fled to Taiwan leaving behind a crippled economy.

Mao's China

It has become conventional to condemn Mao's China as an economic failure, which ultimately forced the reform era of the 1980s on the government. While there is real substance to this argument, it is worth noting that there were developments during the Maoist period that provide favourable conditions for the eventual economic takeoff after 1978.

Despite an initial accommodation with capitalists, the new economy of Mao's China was established by 1952. Mao's China was always going to have a socialist command economy. Emerging as it did at the start of the Cold War as an ally of the USSR, there was little ideological possibility of China following a different economic model. In addition, the new People's Republic of China (PRC) became part of a post-war Soviet-driven system of economic cooperation, signing trade agreements with most of the newly communist Eastern European countries, and benefiting from Soviet technical assistance. The terms of such assistance were often favourable to China, with Moscow providing steel and factory fittings in return for pork and tobacco. Sometimes the PRC also made gestures of solidarity, as in 1953, when it despatched food to East Germany after East Berlin had been shaken by protests against the government.

Nonetheless, the pressure that shaped China's economy during this period did not come only from one side. The United States made a decision not to recognize the establishment of the PRC in 1949, and along with non-recognition came a trade embargo by the US and its allies. Although other nations, such as Japan, did begin to carry on informal international trade with China, the country still appeared isolated from the non-communist world. In addition, relations with the USSR began to sour from the mid-1950s, and were actively hostile by the mid-1960s. In that context, China's leaders began to think in terms of a siege economy that could be defended in the context of a catastrophe, such as an American-sponsored attempt by Chiang Kaishek to retake the mainland, or a Soviet attack across the northern border.

Many economists condemned this decision in retrospect, arguing that it moved much of China's industry away from the coastal areas from where goods could be more easily exported after production.

However, others have noted that the influx of investment into western China actually improved living standards and provided the basis for later development, such as higher literacy rates, industrial plant, transport infrastructure, and water conservancy projects. In fact, some of these factors have become even more relevant since the 1990s, when the government chose to develop the policy of 'Opening up the West', encouraging migration away from China's overpopulated east to the less-developed western regions. The establishment of the Three Gorges Dam project has also helped to cement the importance of the south-western city of Chongqing, now a regional powerhouse and, at least on paper, China's biggest city in terms of population.

The economy under Mao suffered grievously in many ways, of which the Great Leap Forward is the most notorious. However, in its own terms, it also succeeded in its immediate goals. Unlike Nationalist China, the PRC did not collapse under the weight of its own economic problems. The features for which the PRC economy under Mao has been criticized—import substitution, trying to become self-sufficient, low production of consumer goods—were also factors in a variety of democratic countries during the era from the 1950s to the 1980s, including India and New Zealand. These countries, like China, began to take a different path in the 1980s.

China in the global economy

Economic reform started in the countryside, with farmers given freedom to sell their crops on the free market, and individuals encouraged to set up enterprises (see Chapter 3). In the early 1980s, Deng Xiaoping established the Special Economic Zones (SEZs) in port cities on China's southern coast. This signalled his desire to lay down the first phase of economic growth: it would start with manufacturing and light industry, and would be fuelled by foreign investment which would be tempted in by highly preferential tax rates and labour laws. The opening up of Shanghai in the 1990s provided further impetus to this policy (up to that point, the city had been heavily restricted in the amount of foreign direct investment it was allowed to attract). There was excellent precedent for this strategy, as Deng knew. In the late 19th century, Meiji Japan, forced into breakneck modernization, financed its economic growth by developing export-led growth in areas such as textiles, and only later developing heavy industry. In the aftermath of World War II, Japan and its 'dragon' counterparts, Taiwan, South Korea, and Hong Kong, also rose to economic prosperity on light industry and consumer goods, heavily encouraged by their governments. During the same period, of course, Mao's China had been turning inward.

The export-oriented strategy has been spectacularly successful so far. There were danger points: one of the economic factors that fuelled urban protests in 1988–9 was the rapidly growing inflation rate, which severely reduced the purchasing power of state employees (a situation, had it gone out of control, that would have been frighteningly reminiscent of the hyperinflation that doomed the Nationalists in 1948–9). Yet the situation was brought under control, and in the early 21st century, inflation is generally in single digits, even as consumer spending rises with the creation of a new middle class.

Although state ownership of industries and enterprises is clearly declining, the state and the Chinese Communist Party (CCP) are still heavily entwined with business. Party officials have frequently switched from taking a political role to an entrepreneurial one, and good relations with the CCP are often essential to win licences to do business, or to raise capital to set one up. The state and the party have changed form immeasurably since the era of Mao, but neither has retreated from society; they have merely found new ways to interact with and control it.

The other 'dragons' eventually gave up manufacturing cheap goods, as the countries became richer, wages became higher, and ultimately it became cheaper to move manufacturing to other countries (usually China, in fact). The Chinese labour force is far larger and poorer than that in any other Asian country except India, and it will therefore take longer for its labour costs to price China out of the market, although there are signs that this is beginning to happen in some parts of the country. However, the leadership's plans are already developing for the stage of growth that follows their dominance of the manufacturing market.

For the ultimate ambition of the Chinese leadership is not to be the workshop of the world, producing toys and clothes. From the start of the Four Modernizations, 'science and technology' was one of the key targets for reform, and China's leaders have been well aware that Japan, Taiwan, and Korea soon moved away from becoming manufacturing hubs to providing high value-added goods to the world. In particular, China covets the quality and reputation of Japan's technological expertise. Investment in science, technology, and innovation is now one of the government's top priorities, and total domestic spending on it in 2012 was some 1.98 per cent of GDP, with the aim of spending 2.5 per cent by 2020. Some multinationals are also investing in Chinese research: Microsoft and IBM both have significant basic research laboratories in Beijing.

The global financial crisis of 2008–10 was a very significant moment for Beijing. In the second half of 2008, the collapse of Western economies severely damaged the export markets for Chinese goods, revealing that there had been too much capacity in the economy in the previous few years and that too many goods were chasing too few customers. In November 2008 the government introduced a 4 trillion yuan stimulus package, an amount that was the equivalent of some 14 per cent of 2008 GDP, as well as a monetary expansion policy, for the next two years. At a time of contraction elsewhere, China's economy grew rapidly, with huge spending on infrastructure including airports, railways, and toll roads. By the mid-2010s, the concern was that China was once again suffering from overcapacity, and that a turn to domestic consumption, rather than more building, was needed to keep growth rates up, along with a clampdown on irresponsible lending that came from the sudden influx of money into the economy. Yet there was no doubt that the overall success of China in avoiding the 'credit crunch' and banking crisis that damaged the West gave it new credibility in the eyes of many countries in the global south. It was clear that the model of neoliberal economic modernity which had shaped most Western economies until 2008 was not the only one on offer: instead, a strong state sector and a controlled currency might have their advantages, at least in the short term.

In the 2010s, the greatest challenge for Chinese policymakers will be to manage a lower rate of growth than in the boom years. The economy grew by 7.4 per cent in 2014, a figure that would be the envy of most Western nations, but it was still well below the double-digit rates of the early 2000s. In an attempt to move away from growth fuelled by spending on infrastructure paid for by borrowing, the government has stressed consumption, trying to encourage the Chinese to spend more on consumer goods and services (for instance, by inventing new public holidays that would encourage leisure travel and visits to restaurants and theme parks). As the government pledges to spend more on essentials such as health care, the state provision of which was largely eliminated during the reform period, consumers may be more willing to spend on non-essentials. Yet it is proving a hard task to encourage the Chinese to take their savings from under the mattress and spend them: after all, there have been a large number of rainy days in recent Chinese history.

All of this matters to the rest of the world because China's economy has a central role in the global economy, something that was simply not true at the start of the reform era. For a start, its role as a manufacturing base is clear. In 2014, China's trade surplus with the rest of the world was US$380 billion, a figure that was rising, not falling. China also protects its currency, keeping its value low against other currencies to make sure that Chinese exports are cheap in global terms, and that imported goods remain costly. Under repeated pressure from the US and EU, China has gradually allowed the yuan to appreciate in value, but the People's Bank of China continues to maintain control over the level of that appreciation. However, China has not closed itself behind a protectionist wall. In 2001, it succeeded in a long campaign to join the World Trade Organization (WTO), even though WTO entry would force China to open up its markets for goods and services, and to crack down on intellectual property rights violations. China has observed the path taken by Japan, South Korea, and Taiwan in the post-war era, and has made calculations about how far and how quickly to open its economy.

Internationally, China is becoming ever-more concerned about its own image, because it is making its presence more prominent in Africa and Latin America, where it is becoming an exporter of finance capital (in sharp contrast with its desire to import Foreign Direct Investment (FDI)). In Africa in particular, it has used its influence to invest in countries such as Zambia, Zimbabwe, Nigeria, and South Africa, where minerals, uranium, and oil are found. Chinese investment has often been welcomed in countries such as Zimbabwe with bad human rights records, since that investment has not tended to come with demands on human rights standards (although Western observation of these demands has also been patchy, to say the least). As governments change in Africa, the Chinese may find that having backed a previous unpopular government has made them vulnerable with the new regime. This has led to a greater effort to use public diplomacy to suggest to African populations, as well as leaders, that Chinese investment is an asset for developing countries and that Beijing is not following in the West's colonial footsteps.

China is also using its economic power to create new influence in its own region. In 2015, Beijing announced the establishment of an Asian Infrastructure Investment Bank (AIIB), with $50 billion worth of funds. The aim of the bank is to provide funding for projects in the Asia–Pacific region, thereby providing an alternative to Western-influenced institutions and increasing goodwill toward China. However, it still remains to be seen whether the AIIB will in practice reduce tensions in the region caused by a growing perception that China is seeking dominance, not just influence.

Problems of growth

China is not only sharing some of the successes of the 'dragons' that grew up in Asia during the Cold War. It also suffers from some of the same drawbacks, as well as new problems that were not thought of in the 1960s.

Most noticeable to anyone who visits China is the level of environmental pollution that has come along with growth. This is not entirely a problem of the reform era. Plenty of Mao-era state-run factories poured chemicals into the air and water with abandon. However, the breakneck pace of economic growth since the 1980s made the problem many times worse. Japan in the 1960s, during its period of fastest post-war growth, also had a serious pollution problem. However, there were factors that helped to overcome this (in part at least): the growth of environmental groups in civil society and press exposés of scandals, such as the poisoning of fish with mercury at Minamata, provided some sort of counterpoint to the agenda of government and big business, which were more interested in growing the economy. China lacks a true civil society and environmental campaigners are kept within strict boundaries. The Chinese Ministry of Environmental Protection stated that $230 billion in 2010 of direct losses to the economy had been caused by pollution (around 3 per cent of China's entire gross domestic product).

One of the largest internal drivers to Chinese economic growth is the development of an urban middle class with a high-consuming lifestyle. At the same time, the country's level of consumption is proving an immense strain on resources. A generation ago, Beijing was a city of bicycles. Now it is a warren of endless traffic jams (its public transport system is being expanded but is insufficient for the city's size). In Beijing, the authorities plan to limit the number of cars to 'only' 6 million by 2017 (see Figure 10). The growth of car ownership encapsulates a wider problem: China's energy crisis. Pomeranz's model of the 'great divergence' was predicated in part on Britain's good luck in having access to fossil fuels in the 18th century. The strain on China's supply of such fuel in the 21st century has produced serious restrictions on its potential to grow. Although China has coal supplies, they are of low quality and highly polluting, and its oil supply is limited. This has led it into a much closer partnership with Russia, which supplies China by pipeline, but Russia has other customers and cannot be guaranteed to be reliable in the long term. Nuclear energy is another option, but world supplies of uranium are also limited, although China's new friendliness with African nations, such as South Africa, gives it further access to the raw materials needed. China's desire to move population westward is hobbled by the same problem that the US had in its push west: lack of water. China's hinterland had always been underpopulated for a good reason: it was arid land incapable of producing many crops. Now that official policy is to 'open up the west', the problem of water supply must be solved. Unfortunately, there have been cases of rivers in western China that traditionally flowed downstream into South East Asia being diverted to water western China, leaving parts of Vietnam dry. The prospect of 'water wars', virtual or real, is in the minds of governments around the region. Most of all, China has become the world's largest emitter of carbon into the atmosphere, producing 28 per cent of the world's emissions in 2013. China's middle class is still small compared to its much poorer rural population. When the latter become richer, and when China's very high national savings rate reduces as consumers start buying more, the effects on the global environment could be catastrophic.

10. A woman wears a smog veil as she rides her bicycle in Beijing in 1984. In the years since then, China's environmental pollution has become worse, and it is now the world's largest emitter of carbon dioxide into the atmosphere.

Another persistent problem remains: lack of transparency and the corruption that comes along with it. It is still illegal to reveal 'state economic statistics', including basic economic information that would be public in many economies, and China's new 2015 National Security law includes economic and financial data within its definition of 'security'. The organization Transparency International releases an annual survey of perceived corruption around the world: in 2014, China scored 36 (on a scale of 0 to 100, with 100 being least corrupt), slightly more corrupt than India at 38, but less than Vietnam at 31. China's operation of law is still instrumental rather than principled. Though various aspects of corporate and criminal law have been revised, often with some operational success, the Party still stands above the law, making it hard to operate the 'rule of law' in the classic sense.

Economic modernity?

Is China's economy part of the modern world? In some ways, it seems to reflect assumptions very different from the espousal of free markets and withdrawal of the state that were heard in the West in the 1990s and beyond. The state and party are heavily involved in the Chinese economy, as well as its overseas investments, and the market is still hard for outsiders to penetrate despite China's entry to the WTO in 2001. Corruption and lack of transparency cloud attempts to divine the real story of what is happening in China's economy. Yet China's economy now matters to everyone in the world—in manufactured goods, financial services, and currency and interest rates, China's influence is undeniable. While adapting to the world of the modern, globalized economy, China has also forced the world to reshape its economy, at least in part, to Chinese needs.

Chapter 6
Is Chinese culture modern?

In 1915, the response of radical intellectuals to the fear that president Yuan Shikai would declare himself a new emperor was to propose the idea of a 'new culture'. The same language, but a very different argument, was used by Mao during the Cultural Revolution more than half a century later. The idea that somehow China's culture had contributed to its uneasiness with the modern world has persisted from the time of the Opium Wars up to the present day. Yet contemporary Chinese writers are translated and praised around the world; their films win prizes at international film festivals; and Chinese artists command huge prices at auction in the global art market. The quest for a culture that is simultaneously modern but also derived from Chinese desires and inspirations continues to be at the centre of the Chinese artistic endeavour.

The origins of modern readership

Writing has, from the earliest times, been valued in Chinese culture. For centuries, reading remained a largely elite and male skill. An important part of the modernization of Chinese literature was its engagement with a mass audience and its use of technology to reach that audience. In this process, the Ming dynasty (1368–1644) was important, because it saw an immense rise in the consumption of high and popular culture. China was at peace and became more prosperous, and this fact allowed a market for education, as well as for luxury goods and services, to emerge. In addition, new technology allowed particular types of product to develop, such as woodblock prints and popular novels, which could now be printed in huge numbers. The Jesuit monk Matteo Ricci in the 17th century noted the care and skill involved in creating a woodblock, observing that a skilled printer could make some fifteen hundred copies in a single day.

The high Ming and Qing dynasties saw an interest also in connoisseurship, as well as mass-produced culture. The city of Yangzhou, in particular, became known during the Ming as a centre where 'the people by local custom value scholarship and refinement, and the gentry promote literary production' (see Figure 11). In addition to a love of fine books, calligraphy, and painting, the cultivation of rare plants and acquisition of exotic fruits could be a sign of taste. Such connoisseurship would decline with the economic crisis of the late Qing, and the collective politics and ideologically and militarily driven austerity of much of the 20th century also frowned on the cultivation of such luxurious tastes. Only in the 1990s did the depoliticized, consumerist urban economy in China once again encourage personal collections of art.

The end of the imperial era also proved a watershed for both high and popular culture in China. Perhaps the most notable change in Chinese written culture in the early 20th century was the language reform of the 1910s and 1920s. In the late imperial era, classical Chinese was still used for official documents and works of literature and history. But the spoken language had developed over centuries since the classical form was used, and popular writing, such as novels, plays, and unrespectable writing such as pornography were instead written in a vernacular form of Chinese that reflected the spoken language. By the early 20th century, many reformers felt that it would not be possible for China to progress unless its written language was brought

11. Reproduction of a section of the 'Nine Horses' scroll by Ren Renfa, from the book *Master Gu's Pictorial Album* (1603). During the Ming dynasty, art was reproduced for a wider audience to purchase and appreciate.

into harmony with its spoken form. Hu Shi (1891–1962), who had studied in the US, and then took up a teaching position at Peking University, was particularly instrumental in this movement. Despite China's disunited state in the 1910s, the language reform movement was highly successful. At the time of the 1911 revolution, it was still normal for newspapers and school texts to be in a simplified form of the classical language; by the mid-1920s, it was the *baihua* (vernacular) form which had supplanted it almost entirely. This change also had a major impact on Chinese literature, which was almost entirely written in vernacular Chinese by the late 1920s. It is hard to overestimate the change made by this shift in Chinese culture; the arrival of an officially recognized vernacular form of written Chinese paved the way for mass literacy and the ability of the state to use education and propaganda to engage with its population (most notably under the CCP).

May Fourth and its critics

The interwar years have come to be regarded as the single most important period in the development of modern Chinese literature. The era is sometimes characterized as the 'May Fourth' period, a reference to the patriotic outburst that emerged in protest at the Treaty of Versailles (see Chapter 2), and fuelled a powerful rethinking that became known as the 'New Culture'.

The literature of the New Culture period is notable for its tone of constant crisis. It was the product of a China in flux: the new republic was less than a dozen years old, yet had already fallen victim to warlordism; Chinese society was still racked by poverty; and women and men were trying to find new ways to relate to each other. The writers used their fiction to deal with the pressing problems of a China trying to come into modernity. Many of these authors, including Lao She, Lu Xun, and Qian Zhongshu, lived and worked abroad, and brought global influences into their work.

The writer even now generally considered to be modern China's finest was Lu Xun (1881–1936), the pen-name of Zhou Shuren. Lu Xun's earliest short stories are notable for their savage condemnation of what he saw as the nature of Chinese society: inward-looking, selfish, and self-deluding. In *The True Story of Ah Q*, the anti-hero Ah Q is beaten up by his employers, rejected in love, and eventually wrongly accused of a robbery and executed: but all the time, even on his way to his death, he is convinced that he is making great progress in life. Ah Q is a Chinese everyman, but rather than a sympathetic 'little man' fighting against greater forces, Lu Xun makes it clear that Ah Q is a petty and vainglorious jackanapes, whose problems are of his own making, a clear metaphor for the Chinese people as a whole suffering under warlords and foreign imperialists. During the 1911 revolution, Ah Q's thoughts are all about the revenge that becoming a revolutionary will enable him to take on his enemies:

> All the villagers, the whole lousy lot, would kneel down and plead, 'Ah Q, spare us!' But who would listen to them! The first to die would be Young D and Mr Zhao... I would go straight in and open the cases; silver ingots, foreign coins, foreign calico jackets...

Another famous story, *Diary of a Madman*, points the finger of blame for China's crisis even more explicitly. The story concerns a young man who goes mad and becomes convinced that his friends and family are cannibals who are trying to eat him. Eventually, he consults a book about 'Confucian virtue and morality', only to find that 'between the lines', the real text reads 'eat people'. None too subtly, Lu Xun's message was that China's traditional culture was a cannibalistic, destructive monster. These stories, still read by every school student in China today, cemented Lu Xun's reputation as an anti-Confucian voice for whom the modernization of China could not come too soon. While Lu Xun's later writing takes a more ambivalent tone towards the past, particularly as China's present became more intolerable, he never lost his sense of anger that his country had became engulfed in such difficult times.

On the other hand, Lu Xun's writings are less clear about what form a China which rejected its Confucian past should take. Different visions of modernity are visible in other writers of the time. Mao Dun (1896–1982), the pseudonym of the writer Shen Yanbing, wrote one of the finest evocations of urban modernity in *Midnight* (1933), the opening scene of which describes Shanghai as a city lit up by neon, and the harbour of which is dominated by a huge advertisement beaming out the words 'LIGHT, HEAT, POWER', seemingly describing not just the product advertised, but the city itself. For anyone who has seen the new Shanghai which has grown up since 1990, this will seem a familiar scene: today's skyscrapers and neon lights, absent for more than half a century after the Communists' rise to power, would also strike a chord with anyone who knew Mao Dun's city. Mao Dun's Shanghai was glamorous but ruinous: the protagonists of *Midnight* gamble on the stock market and finally lose everything.

A different aspect of modernity informed the work of Ding Ling (1904–86), the pseudonym of Jiang Bingzhi. Still the most famous woman writer of modern China, Ding Ling made her name with a novella, *The Diary of Miss Sophie* (1927), which discussed in unprecedentedly frank style the sexual longings of a young woman. Even Sophie's name was a gesture towards a modern internationalism: not a usual name for a Chinese, 'Sophie' brought to mind Sofiya Perovskaya, the Russian anarchist revolutionary who tried to assassinate Tsar Nicholas II. Sophie is deeply dissatisfied with her own life, recovering from tuberculosis in Beijing, and is deliberately cruel to her friends to drive them away. Yet she burns with desire for a handsome young man, confiding to her diary: 'I can't control the surges of wild emotion, and I lie on this bed of nails of passion.' Again, the contradictions expressed by Sophie were symbolic of a wider crisis that affected young, urban women in China: new freedoms were open to them, but how were they to make use of them?

Sophie's dilemmas, of course, were very much those of the relatively privileged and educated minority dwelling in China's cities. The plight of the working classes was tackled by another major writer of the era, Lao She (1899–1966), in his novel *Camel Xiangzi* (translated as *Rickshaw*). The protagonist, Xiangzi (an ironic name that literally means 'fortunate') tries to get together enough money for his own rickshaw, but loses it all, along with his fiancée who is forced into prostitution and dies before he can rescue her. The novel ends with a broken Xiangzi picking up cigarette-ends to eke out a miserable living. Lao She's portrayal of Xiangzi is far more sympathetic and humane than Lu Xun's savage caricature of Ah Q. Yet it is clear that Xiangzi, too, is meant to be an 'everyman', and by falling victim to 'individualism', the term Lao She uses to criticize him, he has contributed to his own downfall. Lao She later turned to science fiction to express this anxiety. In his 1933 novella *Cat Country* (*Maocheng*), his protagonist is a space traveller who arrives on Mars and finds that the inhabitants are all cats who spend their time fighting each other, and eventually fall victim to an invasion of tiny people (clearly meant to be the Japanese).

Lao She's extraterrestrial metaphor would have been understood by all the writers of the May Fourth era, who believed that China's great crisis lay in its inability to realize that the nation was in mortal danger. For that reason, the now-classic authors of the May Fourth era read rather gloomily. Perhaps unsurprisingly, although they became well known, their books were not the real bestsellers of the era. That distinction goes to a rather different sort of novel, generically known as 'Mandarin Duck and Butterfly' literature, referring to the traditional romantic fiction that had emerged and been widely circulated in late imperial times. These were escapist fantasies, with often stock characters (such as the martial-arts knight errant) and a limited vocabulary which made them more accessible to a wider readership. Yet these novels, too, changed under the impact of modernity. The most successful author in the genre was Zhang Henshui (1895–1967). His novel *Shanghai Express* (1935) draws on traditional fiction in its breezy, popular style and form. Yet its characters are taken from the real, changing China of the 1920s: a 'new woman' dressed up in fancy Western clothes, a business tycoon, and a teacher, among others. Most importantly, it is set on a train, a powerful symbol of modernity, speed, and progress. His biggest hit, however, was the novel *Fate in Tears and Laughter* (1930), a long and picaresque tale which narrated the decision of the hero to choose between two girlfriends, a traditional drum-singer and a Westernized bureaucrat's daughter, who insists on being called 'Miss Helena', English-style. The novel is full of hair-raising escapes, martial arts, and wild romance, yet its central theme, the choice between tradition and modernity, is clear.

Artists as well as writers tried out the new modern techniques of composition, often combining traditional Chinese art forms (such as the landscape) with modern themes. Perhaps the most famous of the artists to work in this hybrid style was Xu Beihong (1895–1953), but other artists used techniques such as the modernist woodcut to develop a spare new style; the artist Feng Zikai (1898–1975) became particularly noted for his skill in this genre. Yet art had also been a commercial enterprise for centuries: from the Ming to the late Qing, the techniques of reproduction that had allowed mass printing of books had also enabled visual images to be produced and sold for a market beyond the elites, and this market persisted and grew into the 20th century.

Writers and artists under Mao and reform

The defining terms of artwork under Mao were set during the war years, when in 1943, he delivered the 'Yan'an Talks on Art and Literature'. Mao made it clear that, in the communist China that he envisaged, 'Literature and art are subordinate to politics...It is therefore a particularly important task to help [artists and writers] overcome their shortcomings and win them over to the front that serves the masses of workers, peasants and soldiers.' With the establishment of the People's Republic in 1949, a brief era of Soviet modernism took hold in China. Visual artists during the Mao years followed prescribed styles such as Soviet-influenced Socialist Realism, and art drawn from folk traditions. The post-1949 years produced few novels of real distinction; some key figures of the May Fourth era such as Lao She and Shen Congwen found it easier to write little or nothing at all. A brief period of openness came with the Hundred Flowers movement in 1957, when authors were given official permission to write as they saw fit. However, Mao became alarmed at the strong criticisms that were voiced and rapidly clamped down by starting the Anti-Rightist Campaign, which flushed out hidden 'traitors' who had supposedly used the opportunity of openness to attack the Party. Many purged in the Campaign, such as Ding Ling, were exiled for over a decade to the far north-east of China.

A new period of creativity opened up in the 1980s and has continued since then (with a difficult period following Tian'anmen in 1989). The contemporary literary scene in China operates in a grey zone: many subjects are still taboo, but, as in the May Fourth era, there is a certain amount of space for critical writing. Many of the best-known authors write with a jaundiced eye about modern China. Mo Yan (the pen-name of Guan Moye) (1955–) has carved out a reputation as one of China's major contemporary novelists, a status cemented by his winning the Nobel Prize for Literature in 2012. His books include *The Garlic Ballads*, *The Republic of Wine*, and *Big Breasts and Wide Hips*, the latter of which attracted criticism because of its explicit sexuality and its lack of moral distinction between Communists and Nationalists during the Civil War (the book was for a time withdrawn in China after selling some 30,000 copies). Wang Shuo (1958–) is another author who successfully negotiated the grey zones of cultural production in China with novels such as *Please Don't Call Me Human* and

Playing for Thrills. Wang was a 1990s bestseller, with over twenty novels published in China, and a national reputation as a major writer. Yet his work was characterized as 'hooligan literature' (*pizi wenxue*) for its nihilistic style and themes. The experiences of both writers show the ambiguities in contemporary censorship. It is quite common, as with Mo Yan's novel, for a book to be released officially, only to be banned later; or, as with much of Wang Shuo's work, for books to be condemned without ever officially being banned. At the same time, both Mo Yan and Wang Shuo continued to draw state salaries and were interviewed in the official press. The fact that these authors can publish daring work is in part a consequence of China's decision to open up again to the outside world. Wang Shuo and Mo Yan are now well-known international literary figures, and this has meant that even when their work is suppressed in China, they themselves remain free. Not all authors are anything like so fortunate, but the boundaries of censorship in China are flexible, nonetheless.

Along with writers, fine artists and musicians have negotiated a new bargain with the state since 1978. Their freedom to paint or play what they wish is much greater than under Mao. On the other hand, orchestras and artists were guaranteed a state income, like all other employees. Now, they must operate under the same commercial constraints as any other entrepreneurs. Two of the most feted contemporary artists have both enjoyed commercial success but under very different circumstances. Ai Weiwei (1957–) is one of China's major contemporary artists, whose sculptures such as *Sunflower Seeds* (2010) are shown at major venues such as London's Tate Modern. Ai's political activism in favour of greater democracy and transparency has brought him into increasing confrontation with the Chinese authorities, with whom he has been in a cat-and-mouse situation since 2011, and who have brought against him a variety of charges including tax evasion. In contrast, Xu Bing (1955–) makes no ostensible statements about politics at all. Nonetheless, works such as *Book from the Sky* (1987–91), which uses plausible-looking but actually non-existent

Chinese characters, hints at the instability of language and meaning in a way that casts doubt on the brassy political certainties of contemporary China.

Moving pictures

The 20th century also heralded a significant change in the way in which the Chinese told stories: film, and later, television. Film came swiftly to China, and by 1927, there were already over 100 cinemas in the country (the majority in Shanghai, the exemplar of Chinese modernity). Hollywood movies were immensely popular in the 1930s, but the Chinese also developed a powerful indigenous industry, again mainly centred on Shanghai. The wartime years reflected the splits within China itself, with patriotic films being produced in the Nationalist areas, while film-makers in Shanghai and Manchuria worked under Japanese occupation. However, like their French counterparts during the same period, it is possible to see hidden resistance to occupation within the latter films. After the victory over Japan, film-makers also reflected the ambiguity of victory. In the 1947 film, *A Spring River Flows East*, the narrative unfolds to show that a family separated by war found that there was no happy reunion, as personal betrayals by family members echo wider ambiguities and betrayals in society about whether it had been better to flee westward in 1937 or stay behind under occupation.

Film-making under Mao mostly reflected the propaganda requirements of the regime, and in the Cultural Revolution period, very few films were made at all. The reform era of the 1980s also marked the beginning of a new, powerful cinema, pioneered by a group who became known as the 'Fifth Generation' of film-makers, and whose film-making should be compared with the simultaneous *glasnost* (openness) film-making of directors such as Elem Klimov which emerged in Gorbachev's Soviet Union. Perhaps its single most prominent exponent was Zhang Yimou (1951–). Although many of Zhang's films were set in the pre-Communist era, and condemned 'feudal' habits such as concubinage, they seemed to reflect an ambiguity about Chinese society in the present day as well, which was a long way from the fervent acclaim for 'New China' that the regime's propagandists had portrayed: the closing frames of *Ju Dou* (1991) shows a chaotic scene at a dyeworks with (symbolically) red paint spilling everywhere, adding to the mess. Scenes such as this led to some of Zhang's later films being banned from distribution within China itself, although they continued to win awards overseas.

Other of Zhang's films, such as *The Story of Qiu Ju* (1992), painted everyday life in rural China in a nuanced and ambiguous fashion (the local official who bullies Qiu Ju's husband by kicking him in the testicles, leading her to try to sue him, also helps her give birth). The contradictions of Chinese modernity, and life under a regime which was unsure of its own identity and purpose, were refracted back through Zhang's pictures. Other directors of the era, including Chen Kaige (1952–), whose films *Yellow Earth* (1984) and *Farewell My Concubine* (1993) also cast a more quizzical eye on the reform era, had a difficult relationship with the censors, though in more recent years many of these directors' films (such as Zhang Yimou's *Hero* (2004)) have included stunning performances of traditional martial arts, using mass media further to publicize aspects of China's traditional culture. However, the boundaries of censorship were tested yet further by underground films; dealing with taboo subjects from homosexuality to tensions during the Cultural Revolution, these films rarely achieve release in China, but can be seen in DVD form at private showings, increasing the scope of the 'grey zone' of culture that is officially banned but still in circulation.

Television only became widespread in China in the 1980s (although a limited service began as early as 1958). However, within a few years, China was rapidly developing the largest television audience in the world, and CCTV (China Central Television) took advantage of the thaw of the reform era to experiment. Many programmes, particularly on news and current affairs, were (and are) weighty and heavy, with positive news intended to boost the party's reputation. More popular were the wide range of costume dramas that filled the screens, including series based on popular classics such as *Outlaws of the Marsh* and *The Romance of the Three Kingdoms*, as well as historical dramas based on both ancient and more recent events, such as the Opium Wars. The power of television, however, was shown most clearly in the debate that surrounded the remarkable television series *River Elegy* (*Heshang*) (see Box 4).

Box 4 *Heshang*

In June 1988, one of the most extraordinary programmes in the history of television was broadcast on CCTV-1, the main Chinese station. It was repeated once (in August 1988). In the aftermath of Tian'anmen Square in 1989 it was banned, and has remained so ever since. The people associated with making it were imprisoned, or fled into exile in Hong Kong or the West.

The programme was called *Heshang* (usually translated as *River Elegy* or *Deathsong of the River*). It was part-documentary, part-polemic. It consisted of six episodes, which reviewed recent Chinese history to try and answer the question of why China was still so backward after a century or more in the modern world. It set out to provoke. Among its main targets were some of the most valued symbols of Chinese civilization: the Great Wall, the dragon, and the Yellow River (the 'river' of the title). Rather than regarding these as symbols of a proud and ancient culture, the film-makers (including writers Su Xiaokang and Wang Luxiang) condemned them as examples of what had held China back: the Wall served to shut China off from the rest of the world; the dragon was a violent and aggressive creature; and the Yellow River was slow-moving, clogged up with silt, and enclosed within Chinese territory. The film-makers made a symbolic contrast with

(continued)

Box 4 Continued

the colour blue, the colour of the Pacific Ocean, from where the new 'spring water' that would renew China will be found. The Pacific Ocean was a not particularly subtle reference to the US, scenes from which were shown throughout the programme. Stirring music accompanied footage of historical events, scientific breakthroughs, and even space exploration.

The programme strongly endorsed the reform programme of CCP General Secretary Zhao Ziyang. Yet it was not merely the expression of one set of views within the leadership. Instead, it had a powerful political agenda of its own, expressed in the line 'Many things in China, it seems, should return to May Fourth'. The May Fourth Movement of 1919, with its catchphrase that China needed 'science and democracy', was at the centre of the film-makers' agenda. However, to say that the promise of May Fourth had not been fulfilled was dangerous indeed. For the CCP itself drew its legitimacy from the fact that it had risen up during the May Fourth era, and that Mao himself had taken part in the intellectual foment of the time. The makers of *Heshang* made their agenda explicit at the very end of episode six: they declared that a dictatorial government was marked by 'secrecy, rule by an individual, and the fickleness of his temperament', whereas democracy was about 'transparency, responsive to popular will, and a scientific approach'. The reference to conservative elements in the CCP, and Mao's legacy, was clear.

The programme created a craze: discussions of it turned up in the newspapers, and letters were sent in their thousands to the television station. Even the bastions of Party orthodoxy such as the *People's Daily* newspaper reprinted parts of the script and printed discussions relating to the series' message.

The events of 3–4 June 1989 ended the thaw in public discussion of which *Heshang* had been one of the most important parts. In autumn 1989, a roundtable of major historians was put together to condemn *Heshang* as reactionary, full of historical errors, and unnecessarily abusive of the Chinese people. In retrospect, the programme does seem to be a historical piece in its almost naïve enthusiasm about the power of Western thought to transform China. Yet the programme was an immensely daring attempt to take on old orthodoxies, but also to use China's past (the May Fourth Movement) to advocate an alternative path. Occasionally, television programmes have shaken an entire society. In the US, *Roots* (1977) forced the country to reconsider its legacy of slavery. In the UK, *Cathy Come Home* (1966) revealed that the lives of Britain's poor could be torn apart by homelessness. For China, there has never been, and may never be again, a television programme as serious, as compelling, as important as *Heshang*.

River Elegy's condemnation of what it regarded as the outdated, inward-looking Chinese civilization daringly included Mao, the 'false peasant emperor', in the list of factors that had held China back. Hundreds of millions of Chinese may well have watched at least some of *River Elegy*'s six episodes before it was banned in the aftermath of Tian'anmen Square in 1989.

Chinese television became more professional and more diverse in the 1990s and early 2000s, although it still remained strongly under the control of the state. It also became more internationalized. One notable example was China's contribution to the global craze for amateur singing competitions (shows such as *Pop Idol* in the UK, and *American Idol* in the US). A television station based in Hunan province launched its own version in 2005, the full title being 'Mongolian Cow Sour Yogurt *SuperGirl* Contest': 120,000 women took part, and the final was viewed by 400 million people, who eventually phoned in to support the 21-year-old Li Yuchun from Sichuan as the winner with 3.5 million votes (see Figure 12).

12. In the centre, Li Yuchun, who won the *SuperGirl* singing contest shown on Chinese television in 2005. Over 400 million people watched, and 8 million sent text messages 'in support' of singers (the word 'vote' was avoided because of its political implications).

The programme was notable for many reasons. In one sense, *SuperGirl* was probably the closest thing that China had had to a free nationwide election since 1912. Certainly it was not well received by the official national broadcaster, China Central Television (CCTV), which declared the programme 'vulgar and manipulative'. The programme was not re-run in its original form the following year, although CCTV's lack of enthusiasm for it may have been fuelled as much by fear for their advertising revenues being sucked away by the regional competitor station as by political worries.

SuperGirl was also notable for how similar it was to other such contests around the world. In the post-Cold War era, the idea was often heard that politics had become almost irrelevant in the Western democracies, as there was little difference between right and left, and citizens had turned instead to materialist consumption and individual gratification. *SuperGirl* suggested that this fear did not just exist in democracies: the idea of individual celebrity as a life goal was as far from the Maoist conception of the good life as could be imagined, and if it distracted young people from political dissent, so much the better, as far as the reform-era state was concerned. In a print advertisement, a line of popular clothing was advertised with the blatantly anti-Maoist slogan 'different from the masses'. The culture of mass-produced individuality was in China to stay.

Architecture and the modern city

Cultural forms changed around the Chinese over the 20th century, and one of the most visible of those changes, indicating the change from premodern to modern, was the way in which city planning changed forever. For centuries, Chinese cities were arranged on a predictable pattern. Travellers approaching a city would first see from miles away the huge, grey-brick walls that surrounded any settlement of size. Within the city walls, the magistrate's *yamen* and bureaucratic offices would be at the centre, and commercial and residential areas would radiate out from that point.

One of the first cities to violate this rule was Shanghai. Expanded as a treaty port whose primary purpose was commerce, not government, the foreign-controlled International Settlement area had a long shopping street, Nanking Road, at its centre, leading not to a government building but to the racecourse. Although the imperialist presence was the source of great anger in China, it was also the example of modernity that was most obviously before people's eyes, and Chinese reformers signalled their own adherence to the norms of the modern city by reproducing them.

So Canton's great walls were destroyed by the Chinese local administration in the 1920s, which was allied to the Nationalist Party (then still a regional rather than national power). In their place came commercial boulevards and highways. The Nationalist government that took power under Chiang Kaishek in 1928 had grand plans for their new capital at Nanjing: it would have arterial roads, tree-lined avenues, electric streetlights, and a palatial new party headquarters that combined features of the Temple of Heaven in Beijing and the US Capitol in Washington. Less than ten years later, the Nationalists had to withdraw from the capital, leaving most of this planning in the realms of the imagination. The war with Japan stopped most of the great building plans, and many cities were heavily bombed, destroying many old buildings and walls.

But the single greatest period of change in China's cityscapes has been the era since 1949. The Nationalists dreamed of using architecture as power: the Communists achieved it. Very few of China's major cities today reflect the topography that the residents of a hundred years ago would have known. All around China, from the 1950s, traditional city walls were knocked down; the winding alleyways of the inner cities were destroyed to make way for high-rise buildings, and old temples and government offices were bulldozed to make way for buildings, some in the Soviet 'gothic wedding-cake' style, but more often in the bland language of the international modern.

Much of the rejection of China's past that marked the Cultural Revolution—literature, philosophy, art—has been reversed since the 1980s, as the Party and the people alike seek to rediscover their own heritage. This reversal has been far less evident in city planning and architecture. To this day, large sections of China's major cities resound with the noise of the wrecking ball and the jackhammer. A significant, if partial, exception is Shanghai. A good proportion of its old colonial architecture, particularly the parts in the city centre, have been carefully preserved, and even sport heritage plaques giving details of their history. Ironically, the indigenous architecture of other cities is under greater threat. The decision to award the 2008 Olympics to the city was a great spur to building in Beijing. Yet the frenzy of building work has almost all been at the expense of the older buildings of the city: the alleyways (*hutong*), the oldest of which appeared in the Yuan dynasty (1271–1368), with their low-rise, courtyard-style living, have been daubed all over the city with the character *chai*, meaning 'for demolition'. The city has replaced them with tower blocks in the suburbs, and sophisticated new buildings in the centre by international architects such as Rem Koolhaas and I. M. Pei. The authorities have given many reasons for the wholesale elimination of so much of old Beijing, citing in particular the unhygienic and impractical nature of the *hutongs* for contemporary living. Many such alleys had no running water in the houses, and the communal toilets were freezing in winter and unbearably close in summer. The old alleys of cities such as Chengdu and Kunming, charming if impractical for motor vehicles and modern plumbing, have also been razed in favour of skyscrapers and tower blocks. But the replacements often show little evidence of a Chinese flavour. In the early 20th century the Nationalists in Canton tried to show their modernity by adapting globalized (that is, Western) urban planning; that tendency is still evident for China's government in the early 21st.

Globalizing Chinese culture?

China exists in a global cultural context. In the 20th century it tended to absorb cultural norms above all, whether of modern literary genres, cinematic styles, or artistic and architectural techniques. However, there are signs that aspects of the trend are reversing and China is beginning to project not just military and economic power, but also cultural strength (sometimes termed 'soft power'). Cinema has helped here, through the popularity of films such as Zhang Yimou's martial arts extravaganzas. Chinese language learning has risen around the world, stimulated by a new perception of China's global importance. Furthermore, one should note the heavy influence of internationalism in the shaping of modern Chinese culture. Even in the most closed period of modern Chinese history, during Mao's rule, cultural models came from the USSR, and Marxist-Leninist ideas were widespread. In the period before 1949 and since 1978, a stunning variety of influences, from American management gurus to French philosophers to Mahatma Gandhi, have reshaped the Chinese sense of the modern self and the meaning of 'Chinese culture'.

Chapter 7
Brave new China?

During the 2010s, China is likely to change yet further. Since the ascension to power of Xi Jinping, China has become less liberal at home and more assertive overseas. Chinese media and academics are subject to increasing restrictions on what they can discuss, and in the South and East China Seas, Chinese claims on territorial and maritime boundaries are becoming more insistent. Yet there are also signs of forward thinking, although often in directions that the liberal world will find uncomfortable. The power of China's currency, the yuan (renminbi), will grow if it moves towards full convertibility and becomes a significant global reserve currency. The Chinese government has proposed a new 'Silk Road' which may bring China closer to central Asia, Iran, and the Middle East. In addition, the Chinese presence in Latin America and Africa continues to grow. Meanwhile, the Party embeds itself further into all aspects of business and social change, particularly since it has adapted to the technology of the 21st century with some aplomb. One thing remains constant, however: the CCP has no intention of allowing any rivals even to think of attaining power in China.

This book started with a 'new China' envisioned a century ago. What you think of modern China may be affected by your response to another vision, not written with China in mind—the modernity of Aldous Huxley's *Brave New World* (1932).

The book's protagonist, the Savage, is brought from the wilds into a 'civilization' set several centuries into the future where everybody is happy: material comforts on demand, everybody slotted into social categories that suit their needs, and dangerous and uncomfortable information kept firmly suppressed. Those who have overactive minds—and they are few in number—end up exiled to Iceland, where the system sends 'all the people who aren't satisfied with orthodoxy, who've got independent ideas of their own'.

Near the climax of the book, the Savage confronts Mustapha Mond, the 'World Controller', who defends the safe, cosy, and unquestioning world that he and his system have created: Mond admits that 'being contented has none of the glamour of a good fight against misfortune…Happiness is never grand'.

The Savage claims 'the right to be unhappy'. The Controller replies:

> 'Not to mention…the right to have too little to eat; the right to be lousy; the right to live in constant apprehension of what may happen tomorrow; the right to catch typhoid; the right to be tortured by unspeakable pains of every kind.'
> There was a long silence.
> 'I claim them all,' said the Savage at last.
> Mustapha Mond shrugged his shoulder. 'You're welcome,' he said.

Of course, both Mond and the Savage are right—and wrong. China today is very far from being a brave new world, even though Shanghai's night cityscape may look like one. But the conversation between the Savage and the Controller says something about the infinitely difficult balancing act that has affected all governments—the Qing of the 'New China' of 1910, the Nationalists, Mao's 'New China' of 1949, or the current leadership's nurturing of 'a Chinese renaissance' (*fuxing*)—in deciding what the relationship will be between the state, the party, and the people in a truly modern China. Can China afford to give people 'the right to be unhappy', or does it need to exile those who ask for it to its own Iceland? Are people who live in desperate poverty able to be free in any meaningful sense? Are those who have television, running water, and a car, but cannot openly discuss their views on politics being infantilized by an over-protective, sometimes vindictive state and party? The answers to those questions are at the heart of the ever-changing, perhaps never-ending, journey to what it means to be modern and to be Chinese.

Timeline

1997	Death of Deng Xiaoping: Jiang Zemin reconfirmed as leader
2001	Beijing awarded the 2008 Olympics
2001	China enters World Trade Organization
2002	Leadership passes to Hu Jintao
2008	Summer Olympic Games held in Beijing
2012	Leadership passes to Xi Jinping
2015	Major commemorations of the 70th anniversary of the end of the War of Resistance against Japan (World War II in China)
2017	19th Congress of the Chinese Communist Party
2022	Beijing to hold Winter Olympic Games

References

Chapter 1: What is modern China?

W. Y. Fullerton and C. E. Wilson, *New China: A Story of Modern Travel* (London, 1910), p. 234.

Chapter 2: The old order and the new

Chen Hongmou: William Rowe, *Saving the World: Chen Hongmou and Elite Consciousness in Eighteenth-Century China* (Stanford, 2001), esp. ch. 9.

Wei Yuan: Philip Kuhn, *Origins of the Modern Chinese State* (Stanford, 2002), pp. 39, 48.

World War I: Xu Guoqi, *China and the Great War: China's Pursuit of a New National Identity* (Cambridge, 2005), esp. Part II.

Chapter 3: Making China modern

Mortality rate in 1930: Lloyd Eastman, 'Nationalist China during the Nanking Decade, 1927–1937', in John K. Fairbank and Albert Feuerwerker (eds), *Cambridge History of China*, volume 13 ('Republican China, 1912–1949'), p. 151.

Madame Chiang Kaishek: Pei-kai Cheng and Michael Lestz, *The Search for Modern China: A Documentary Collection* (New York, 1999), p. 296.

Edgar Snow, *Red Star over China* (orig. 1937; London, 1973), p. 92.

Tan Zhenlin: Jasper Becker, *Hungry Ghosts: China's Secret Famine* (London, 1996), p. 59.

Red Guard quotations: Song Yongyi et al. (eds), *Chinese Cultural Revolution Database* (Hong Kong, 2002).

Wang Hui: Wang Hui, *China's New Order: Politics, Society and Economy in Transition*, ed. Theodore Huters (Cambridge, MA, 2003), p. 180.

Chapter 4: Is Chinese society modern?

Zou Taofen: Rana Mitter, *A Bitter Revolution: China's Struggle with the Modern World* (Oxford, 2004), p. 69.

Li Yu: Dorothy Ko, *Cinderella's Sisters: A Revisionist History of Footbinding* (Berkeley, 2005), p. 152.

Mao on Miss Zhao: Stuart Schram (ed.), *Mao's Road to Power* (Armonk, NY, 1992–), vol. 1: 423.

Ageing: Michael Backman, *The Asian Insider* (Houndmills, 2006), p. 225.

Figures on incomes: Jonathan Kaiman, 'China gets richer but more unequal,' http://www.theguardian.com/world/2014/jul/28/china-more-unequal-richer; data based on Peking University Institute of Social Science survey data.

Chapter 5: Is China's economy modern?

China's economy bigger than the US: Ben Carter, 'Is China's economy really the largest in the world?' at http://www.bbc.co.uk/news/magazine-30483762.

Brandt: Loren Brandt, *Commercialization and Agricultural Development: Central and Eastern China, 1870–1937* (Cambridge, 1989), ch. 7.

Economy of western China: Chris Bramall, *In Praise of Maoist Economic Planning: Living Standards and Economic Development in Sichuan since 1931* (Oxford, 1993), esp. pp. 335–40.

2008 financial crisis: Yu Yongding, 'China's response to the global financial crisis,' *East Asia Forum* (24 January 2010), http://www.eastasiaforum.org/2010/01/24/chinas-response-to-the-global-financial-crisis/.

Cost of pollution: Edward Wong, 'Cost of Environmental Damage in China Growing Rapidly Amid Industrialization,' (*New York Times*, 30 March 2013), http://www.nytimes.com/2013/03/30/world/asia/cost-of-environmental-degradation-in-china-is-growing.html?_r=0.

Chapter 6: Is Chinese culture modern?

Yangzhou: Timothy Brook, *The Confusions of Pleasure: Commerce and Culture in Ming China* (Berkeley, 1998), p. 128.

Nine Horses scroll: (picture) Craig Clunas, *Art in China* (Oxford, 1997), p. 182.

Ah Q: 'The True Story of Ah Q', in Lu Xun, *Call to Arms*, tr. Yang Xianyi and Gladys Yang (Beijing, 1981), p. 99.

Yan'an Talks: Bonnie S. Macdougall, *Mao Zedong's 'Talks at the Yan'an conference on Literature and Art': A Translation of the 1943 Text with Commentary* (Ann Arbor, 1980).

Zhang Xiaogang: Jonathan Watts, 'Once hated, now feted', *Guardian*, 11 April 2007, p. 29.

Nanjing city planning: William Kirby, 'Engineering China: Birth of the Developmental State', in Wen-hsin Yeh (ed.), *Becoming Chinese: Passages to Modernity and Beyond* (Berkeley, 2000), pp. 139–41.

Chapter 7: Brave new China?

Aldous Huxley, *Brave New World* (orig. 1932; London, pbk, 1984), pp. 178, 192.

Further reading and listening

The books on this list are mostly academic works, but I have deliberately included here a number that have been written with at least a partially non-academic readership in mind, and which do not demand a comprehensive knowledge of China to be read profitably. No slight is intended to the many colleagues whose more specialist monographs and articles I have drawn on to compose this volume.

Online listening

Three BBC radio documentaries made by the author on topics covered in this book:

China's New Iron Rice Bowl: on changing ideas of social welfare and economic change http://www.bbc.co.uk/programmes/b01mqpgt

Shanghai: World City Redux: on why Shanghai is rediscovering its cosmopolitan past as it shapes its future http://www.bbc.co.uk/programmes/b03rx8jf

Enter the Dragon: the world of cutting-edge drama in China today http://www.bbc.co.uk/programmes/b04lpqnj

The pattern of modern Chinese history

John Fairbank et al. (eds), *The Cambridge History of China*, vols 10–15 (Cambridge, 1978–): detailed essays summarizing key themes in political, cultural, and social history.

Jonathan Spence, *The Search for Modern China* (New York, 2013): comprehensive survey history from the 17th century to the present.

Jeffrey Wasserstrom, *China in the 21st Century: What Everyone Needs to Know* (Oxford, 2nd edn, 2013): incisive account of key issues facing China as it goes through the growing pains of a superpower in the making.

Odd Arne Westad, *Restless Empire: China and the World Since 1750* (London, New York, 2013): an account of China's foreign relations packed with information and insight.

Pre-1949 China

Robert Bickers, *Empire Made Me: An Englishman Adrift in Shanghai* (London, 2003): gripping account of the complexities of imperialism in China, told through one man's life.

Lloyd Eastman, *Family, Fields, and Ancestors: Constancy and Change in China's Social and Economic History, 1550–1949* (New York, 1988): accessible guide to social and economic change in China from the Ming to 1949.

Henrietta Harrison, *The Man Awakened from Dreams: One Man's Life in a North China Village, 1857–1942* (Stanford, 2005): moving portrait of one man living in rural China from the late Qing to the war against Japan.

Rana Mitter, *China's War with Japan, 1937–45: The Struggle for Survival* [US title: *Forgotten Ally*] (London, Boston, 2013): an account of China's experiences in World War II that led to millions of deaths and paved the way for revolution.

Kenneth Pomeranz, *The Great Divergence: China, Europe, and the Making of the Modern World Economy* (Princeton, 2000): highly influential study of the differences between the economic development of China and Europe.

Philip Short, *Mao: A Life* (London, 1999): deeply researched and thoughtful study of the life of Mao Zedong.

Politics, society, and culture in the reform era

Geremie Barme, *In the Red: On Contemporary Chinese Culture* (New York, 1999): wide-ranging account of China's cultural scene by an expert observer.

Richard Baum, *Burying Mao: Chinese Politics in the Age of Deng Xiaoping* (Princeton, 1994): detailed and clear account of China's development from 1976 to the mid-1990s.

Joseph Fewsmith, *China since Tiananmen: The Politics of Transition* (Cambridge, 2001): particularly strong on intellectual debates in China.

Karl Gerth, *As China Goes, So Goes the World: How Chinese Consumers are Changing Everything* (New York, 2013): fascinating account of how consumer choices affect economic, politics and environment in China and beyond.

Robert Gifford, *China Road: One Man's Journey into the Heart of Modern China* (London, 2008): insightful and often poignant reflections on change in modern China from NPR's expert former China correspondent.

Graham Hutchings, *Modern China: Companion to a Rising Power* (London, 2000): comprehensive handbook dealing with a wide range of historical and contemporary topics.

Richard Curt Kraus, *The Party and the Arty: The New Politics of Culture* (Lanham, MD, 2004): clear explanation of how China's art and cultural world has entered the commercial era.

Norman Stockman, *Understanding Chinese Society* (Cambridge, 2000): detailed introduction to changes and continuities in Chinese social structures.

Index

Index

Modern China